INVERSE PROBLEMS

Teachers may reproduce these projects for their students. However, the projects remain the property of The Mathematical Association of America and may not be used for commercial gain.

This project was supported, in part, by the National Science Foundation. Opinions expressed are those of the author and not necessarily those of the Foundation.

Library of Congress Catalog Card Number 99-62793

ISBN 0-88385-716-2

Printed in the United States of America

Current Printing (last digit):
10 9 8 7 6 5 4 3 2 1

INVERSE PROBLEMS

Activities for Undergraduates

C. W. Groetsch
Department of Mathematical Sciences
University of Cincinnati

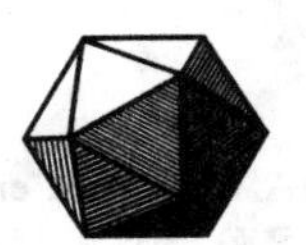

Published and Distributed by
THE MATHEMATICAL ASSOCIATION OF AMERICA

Classroom Resource Materials is intended to provide supplementary classroom material for students—laboratory exercises, projects, historical information, textbooks with unusual approaches for presenting mathematical ideas, career information, etc.

101 Careers in Mathematics, edited by Andrew Sterrett
Calculus Mysteries and Thrillers, R. Grant Woods
Combinatorics: A Problem Oriented Approach, Daniel A. Marcus
A Course in Mathematical Modeling, Douglas Mooney and Randall Swift
Elementary Mathematical Models, Dan Kalman
Interdisciplinary Lively Application Projects, edited by Chris Arney
Inverse Problems, by C. W. Groetsch
Laboratory Experiences in Group Theory, Ellen Maycock Parker
Learn from the Masters, Frank Swetz, John Fauvel, Otto Bekken, Bengt Johansson, and Victor Katz
Mathematical Modeling for the Environment, Charles Hadlock
A Primer of Abstract Mathematics, Robert B. Ash
Proofs Without Words, Roger B. Nelsen
A Radical Approach to Real Analysis, David M. Bressoud
She Does Math!, edited by Marla Parker

MAA Service Center
P. O. Box 91112
Washington, DC 20090-1112
1-800-331-1622 fax: 1-301-206-9789

Preface

This is not a textbook. Nor is it a survey of elementary inverse problems. It *is* a personal miscellany of activities related to inverse problems that is meant to enrich, and perhaps enliven, the teaching of mathematics in the first two undergraduate years.

The first, and one of the most difficult, hurdles to be cleared in discussing inverse problems is the definition of an inverse problem. Justice Potter Stewart, when speaking of pornography, said that he couldn't define it, but he knew it when he saw it. The Supreme Court eventually ruled that pornography is defined by "community standards." In the same way, mathematicians find it difficult to define "inverse problem," yet most recognize one when they see it. This recognition comes from a vague familiarity with accepted scientific "community standards" for *direct* problems that have been developed over the course of history. With this in mind, I have found that an indirect approach, one consisting of historical vignettes on inverse problems, is an effective way to introduce these problems. The first chapter consists of a number of such stories, any or all of which may be omitted (but I think this would be a mistake).

Following this are modules on inverse problems in precalculus, calculus, differential equations, and linear algebra. Each module consists of a brief introduction, a collection of "Activities," and some notes with suggestions for further reading. All modules are prefaced with advice on course level, mathematical and scientific background required, and technology needed to perform some Activities. The introductions are aimed at the teacher's level, not that of the student. This guide is meant to be a resource for *teachers*, not a collection of materials to be handed directly to students. Teachers are invited to choose the ideas they like, select (and perhaps modify) activities that they feel are appropriate for their students, and use them as they see fit in their courses. In particular, there is no necessity to use a module as a whole.

The Activities are of six kinds. *Questions* are just that and are meant to be answered in a narrative style. *Exercises* are routine problems, while *Problems* present a bit more (sometimes quite a bit more) of a challenge. A *Calculation* requires the use of a graphical–symbolics calculator, such as a TI-92 or equivalent. A digital computer and appropriate software are required for a *Computation*. Scripts in MATLAB®, the tool of choice for many in the inverse problems community, are provided (MATLAB is a registered trademark of TheMathWorks, Inc.). The algorithms in these scripts are simple enough that readers who have formed attachments to other software should not find translating the codes unduly burdensome. The scripts may be downloaded from the author's web page (linked through http://math.uc.edu). Finally, *Projects* are open-ended activities that allow broad scope for student effort and imagination.

I am indebted to many anonymous readers who made helpful comments and suggestions for improving the (at times, seemingly endless) series of preliminary editions of this work. In particular, the final product owes much to the sharp eyes of Robin Endelman, Terry Sullivan, and Zongjun Zhang. Lee Zia provided much-appreciated encouragement throughout the course of this project and I benefited from helpful comments by Gunther Uhlmann and Graham Gladwell. This work was supported by a grant from the National Science Foundation.

Contents

1

Introduction to Inverse Problems

1.1 What Are Inverse Problems?

> Usually in mathematics you have an equation and you want to find a solution. Here you were given a solution and you had to find the equation. I liked that.
>
> Julia Robinson

Inverse problems are hard to define. Yet nearly every mathematician recognizes an inverse problem when she sees one. As children we learn about the **direct** problem of multiplication: given two numbers we find their product. The corresponding **inverse** problem is to find a pair of factors of a given number. We soon learn that, like many inverse problems, the factorization problem does not always have a unique solution. In fact, it is in trying to impose uniqueness on the solution of this inverse problem that we are led to the notion of prime numbers, and from this a whole world of mathematical possibilities opens up.

What is the oldest problem in mathematics? I like to think that it is the linear interpolation problem. Herodotus traced the origins of geometry to a guild of linear interpolators: the rope stretchers of ancient Egypt. Today we would call the linear interpolation problem an inverse problem (note that this inverse problem may have a unique solution, no solution, or infinitely many solutions, depending on the number and nature of the points). The direct problem is to calculate the values of a given linear function; in the inverse problem we must determine a linear function from a couple of points on its graph. This, in essense, is what the rope-stretchers did: They solved, in a very direct way, an inverse problem.

Inverse problems come paired with direct problems and of course the choice of which problem is called *direct* and which is called *inverse* is, strictly speaking, arbitrary. However, the academic mathematical community has embraced what are now called direct problems with a warmth not generally extended to inverse problems. The greater part of undergraduate training in mathematics is dominated by *direct* problems, that is, problems that we can characterize as those in which exactly enough information is provided to the student to carry out a well-defined stable process leading to a unique solution. Typically, a process is described in detail, and an appropriate input is supplied to the student, who is then expected to find the unique output. In the sciences, the process is usually called a model, with the input labeled the cause and the output the effect. The prevailing paradigm in direct problems may therefore be described by the following:

$$\text{input} \longrightarrow \text{process} \longrightarrow \text{output}$$
$$\text{or}$$
$$\text{cause} \longrightarrow \text{model} \longrightarrow \text{effect.}$$

If we symbolize the input by x and the process by K, then the direct problem is to find Kx, the value of an operator at a point in its domain. The direct problem may therefore be portrayed schematically as follows:

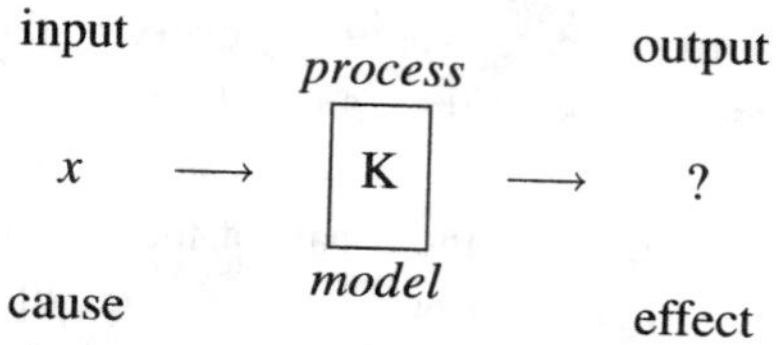

input	*process*	output
$x \longrightarrow$	K	$\longrightarrow ?$
cause	*model*	effect

But clearly this is only one-third of the story. Two inverse problems are immediately suggested by every direct problem. One is the *causation* problem: given a model K and an effect y, find the cause of the effect. This inverse problem is described by the following:

input	*process*	output
$? \longrightarrow$	K	$\longrightarrow y$
cause	*model*	effect

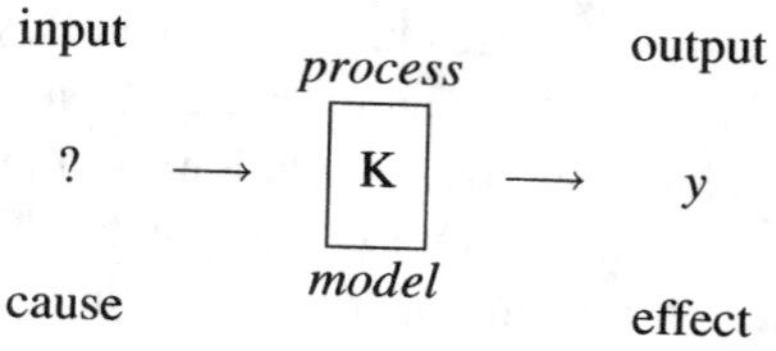

Of course, the causation problem for K is the direct problem for K^{-1}, assuming that the model is invertible, but for the most part we will assume that in the direct problem the model is not necessarily invertible. It is in this sense, and also at times in the sense of historical precedence, that we will distinguish between the direct problem and the inverse causation problem. In the elementary example discussed above, we consider the multiplication problem to be the direct problem and the factorization problem to be the inverse problem. Note that historically the multiplication problem probably came first and that the process of multiplication is not uniquely invertible.

The other inverse problem suggested by the direct problem is the *model identification* problem: given cause–effect information, identify the model. This inverse problem is illustrated in the following diagram:

input	*process*	output
$x \longrightarrow$	?	$\longrightarrow y$
cause	*model*	effect

If the process K is truly an operator, that is, a *function*, then for any given input in its domain, a unique output is determined. That is, the direct problem has a unique solution. On the other hand, there is no guarantee that the inverse causation and model identification problems have unique solutions. Furthermore, if the operator K is continuous in some sense, then the solution of the direct problem is stable with respect to small changes in the input (these changes are of course gauged by the topologies in the domain and range spaces of the operator). Even when the operator has a well-defined inverse, so that the causation problem is uniquely solvable, there is no guarantee that the solution of the inverse problem is stable; the inverse operator may in fact be discontinuous.

Inverse problems have been enormously influential in the development of the natural sciences—a fact not generally appreciated. The conventional approach to the teaching of the natural sciences has emphasized direct problems. Given causes and models describing the evolution of causes into effects, the problem is to find the effects. In this scheme of things the dominant viewpoint is direct, future-oriented, and outward-looking. Most problems involve prediction or the determination of external characteristics of a known internal cause. However, great advances in science and technology have been made possible by solving inverse problems. Such problems involve determining physical laws through indirect observations, remote sensing, indirect measurement, finding the nature of an inaccessible region from measurements on the boundary, the reconstruction of past events from observation of the present state, and many others. In such problems the attitude is indirect, past-directed, or introspective and the problems often involve "postdiction," noninvasive evaluation, or nondestructive testing.

The best way to gain some insight into the nature of inverse problems and the modes of thought that lead to their posing is to see some examples of inverse problems. This could be done with contrived mathematical examples, but I have found that a historical approach is a more effective way of introducing inverse problems. The remainder of this chapter consists of historical vignettes of important or interesting inverse problems, some of which have changed the way in which we think about our world. There is no mathematics in this chapter, but there is a great deal of mathematics underlying it. It is not necessary to read any of the vignettes before proceeding to the modules in this guide, but I strongly recommend that the reader at least skim a few of the sections in order to appreciate the historical role of inverse problems in the sciences. When studying any of the vignettes, the reader's goal should be to differentiate between the direct and inverse problems and consider the implications of each.

1.2 Archimedes' Bath

We enter the future backwards.
P. Valéry

In Book VII of his *Republic*, Plato (428? B.C.–348? B.C.) discusses a situation that illustrates perfectly some of the issues raised by inverse problems. Plato sets the scene:

> Behold! human beings living in an underground den; here they have been from their childhood, and have their legs and necks chained so that they cannot move, and can see only before them, being prevented by the chains from turning round their heads. Above and behind them a fire is blazing at a distance, and between the fire and the prisoners there is a raised way; and you will see, if you look, a low wall built along the way, like the screen which marionette players have in front of them, over which they show their puppets.

In Plato's story, the captives are faced with reconstructing the real world outside of the cave on the basis of very limited information—observations of shadows projected on the back of the cave. That is, they seek the cause (real objects) of the effects (projected shadows) of the distant fire (the model). The

direct problem is completely understood: given an object on the wall, it is a routine matter, knowing the process by which the fire casts the shadow of an object, to completely specify the unique shadow that a given object casts. On the other hand, the **inverse** problem of determining an object from its shadow does not have a unique solution. For example, a square image cast on the back of the cave may correspond to a cube or a right circular cylinder with equal height and diameter, or in fact infinitely many other three-dimensional objects. Furthermore, shadows that are very nearly the same may correspond to three-dimensional objects the difference of whose volumes is arbitrarily large, that is, the inverse problem is in a certain sense *unstable*. In the problem of the cave, the model, that is, the projecting property of the fire, destroys information irrevocably—an entire spatial dimension is suppressed. In mathematical terms, we would say that the operator has a nontrivial null-space and hence that the data for the inverse problems, that is, the shadows, lack essential information necessary to uniquely reconstruct the object. As we shall see, this is a common feature of many inverse problems.

The sphericity of the earth has been accepted from ancient times. Pythagoras held that the earth was spherical in part because the sphere is the "most perfect" figure. Plato's student Aristotle (384–322 B.C.) presented several arguments for the sphericity of the earth, including one based on inverse theory. Aristotle says

> How else would eclipses of the moon show segments shaped as we see them? As it is, the shapes which the moon itself each month shows are of every kind—straight, gibbous, and concave—but in eclipses the outline is always curved: and, since it is the interposition of the earth that makes the eclipse, the form of this line will be caused by the form of the earth's surface, which is therefore spherical.

Thus the shape is not obtained by *direct* observation, but rather by *indirect* reasoning from observations of the earth's shadow on the moon's surface. This was Plato's cave problem applied to the heavens and may be considered the first step in a field of mathematics known today as *geometric tomography*.

However, the locally flat nature of the earth made it clear that the circumference of the sphere must be very large indeed. Plato in his *Phaedo* says "I believe the earth is very large and that we ... live in a small part of it about the sea, like ants or frogs about a pond." Although he doesn't mention how the estimate was obtained, Archimedes, in *The Sand-Reckoner*, says that his predecessors (including, presumably, his father the astronomer Pheidias) took the circumference of the earth to be 300,000 stadia, or over 34,000 miles. Clearly it was impossible to measure the circumference of the earth *directly*, say by pacing it off. It was Eratosthenes of Cyrene (276 B.C.–194 B.C.) who made an *indirect* attack on the problem. If an arc on a great circle of the earth and the circumference of the earth could be directly measured, then the angle subtending the arc could be obtained by simple proportion—a routine **direct** problem. Eratosthenes devised an indirect method for measuring the angle and solved the **inverse** problem of determining the circumference of the earth from the measure of the angle and the length of the arc subtending the angle. His estimate was derived using his knowledge of the following facts: Alexandria and Syene lie on the same meridian, 500 miles apart, and Syene is located on the extremity of the summer tropic zone, that is, on the Tropic of Cancer. The situation of Syene could be verified by noting that at midday on the summer solstice the sun was directly overhead in Syene. On the other hand, at the same time and date a vertical rod placed in the ground at Alexandria cast a shadow at an angle sweeping out an arc equal to one-fiftieth part of a full circle. The circumference of the earth is therefore about fifty times the distance from Alexandria to Syene, or about 25,000 miles. Although Eratosthenes' assumptions are not exactly true, and the figure he took for the distance from Alexandria to Syene is a matter of some controversy, his argument is a classic example of inverse reasoning applied to estimate an inaccessible quantity.

Our final illustration of an ancient inverse problem is the famous story of Archimedes' bath. King Hieron of Syracuse had comissioned a new gold crown and suspected that the goldsmith had cheated him by adulterating the gold in the crown with a lighter metal. There was of course a *direct* method to test his suspicion: simply melt down the crown to determine its volume and compare its weight to an equal volume of pure gold. But it seemed a shame to destroy such a beautiful crown, so Hieron asked Archimedes (287 B.C.–212 B.C.) to devise an *indirect* method to test the gold content of the crown—what is called in today's technical literature a *nondestructive evaluation* (NDE) technique. The idea for the earliest NDE technique came to Archimedes one day while he was taking a bath. As Plutarch tells it:

> According to the story, Archimedes, as he was washing, thought of a way to compute the proportion of gold in King Hieron's crown by observing how much water flowed over the bathing stool. He leapt up as one possessed, crying *Eureka!* ("I've found it!"). After repeating this several times, he went his way.

Archimedes' indirect NDE scheme is as simple as it is ingenious. Use a balance to weigh out a lump of gold of weight equal to that of the crown. Immerse the lump in water and measure the volume displaced. If that volume is smaller than the displacement of the crown, then the crown was adulterated with a lighter metal. It is not certain that this was actually the method Archimedes used, as the difference in volume could be exceedingly small. Another possible approach that Archimedes might have used is to carefully balance the crown with an equal weight of gold and then immerse the balance, crown, and gold in water to note any imbalance in favor of the pure gold sample. In any case, tradition has it that the goldsmith was found guilty of fraud, presumably with unpleasant consequences.

1.3 Tartaglia's Wager

> In fact, military engineering in all its forms ...
> was the main field of technical invention
> throughout the Renaissance.
> George Sarton

In Renaissance Italy advancement was swift for those who could provide useful information to the local ruling family—and few things were more useful to the Italian princes than gunnery. Good gunnery could be the decisive factor in disputes with other principalities, and it was vitally important in resisting the great extra-European threat of the time: the Ottoman Empire. The self-taught Brescian mathematician Nicolo Tartaglia (Nicolo Fontana, 1500?–1557) is chiefly remembered today for his algebraic duels with Antonio Fiore and Girolamo Cardano (1501–1576), in which the weapons of choice were algorithms for solving cubic equations. But Tartaglia was a great innovator in science as well. His book "The Newly Discovered Invention of Nicolo Tartaglia of Brescia, Most Useful for Every Theoretical Mathematician, Bombardier, and Others, Entitled *New Science*," dedicated to the Duke of Urbino and published in 1537 in response to the threats of Suleiman the Magnificent, was the first serious attempt to provide a mathematical basis for ballistics. Unlike Galileo a century later, Tartaglia was unable to completely escape Aristotle's grip and therefore his description of trajectories was seriously flawed. He did however address two important inverse problems in gunnery.

Tartaglia claimed to be the inventor of the gunner's square. This was a carpenter's square with one long arm which was inserted into the barrel of a cannon. Attached across the right angle of the square was a protractor graduated into 12 "punti" or points that were read off by a plumb line attached to the vertex of the right angle (incidentally, a horizontal shot would correspond to a blank reading on the gunner's square, hence the term "point blank"). Using Tartaglia's square, a gunner could measure the elevation in "punti" of his field piece, make a test firing, and measure the resulting range of the shot. This may be considered a **direct** problem. Tartaglia was interested in the **inverse** problem: given a desired range, find an angle of elevation to achieve that range. In particular, Tartaglia, spurred by inquiries from some Veronese gunners in 1531, was interested in the elevation angle that produces the maximum range, and he claimed that this occurs at the sixth point of the gunner's square, i.e., at 45°. This issue of the optimal angle was a matter of some dispute and was the subject of a wager between Tartaglia and the Veronese gunners. It is interesting that the military men thought that the 45° aiming was a bit too high (more on this later; see the module *It's a Drag*). In studying the inverse problem, Tartaglia discovered a surprising fact. In his words, "I knew that a cannon could strike in the same place with two elevations or aimings, and I found a way to bring this about, a thing not heard of and not thought by any other, ancient or modern." That is, the inverse problem had nonunique (in fact, exactly two) solutions.

About a century after Tartaglia, Galileo Galilei (1564–1642) put the subject of elementary ballistics in a vacuum on a firm mathematical basis. His law of inertia and law of falling bodies were the key ideas he used in establishing that the trajectory is parabolic and in providing rigorous proofs of Tartaglia's assertions concerning the inverse problem. This work of Galileo's is generally taken to be the origin of mathematical physics.

Interest in ballistics of course continued (and continues), but we will close the early history of this topic by mentioning an inverse problem studied by Edmond Halley (1656–1742), of comet fame, in 1686. Halley noted that it would be useful to find a way to hit a given target while using as little powder charge as possible. This would of course save powder, but Halley also remarked that cannon balls fired with too much force tend to "bury themselves too deep in the ground, to do all the damage that they might ... which is a thing acknowledged by the besieged in all towns, who unpave their streets, to let the bombs bury themselves, and thereby stifle the force of their splinters." Halley's inverse problem was to find a way to strike the target while using a minimal charge. Using geometric reasoning, he gave a clever solution to this inverse problem.

Interest in direct and inverse problems for trajectories continues in modern times, particularly during times of conflict and crisis. According to RAF Group Captain W. W. Wintherbotham, the solution by R. V. Jones of an inverse trajectory problem involving V1 buzz bombs during World War II was crucial in identifying V1 launch sites for attack:

> From the plots, Jones was able to work out the dispositions of all the plotting stations and obtain detailed performance figures on most of the flying bomb

> trials. When he had worked out the coordinates of the launching point at Peenemunde, by *backtracking* from the radar plots, he asked for a spy-plane sortie over that pinpoint and also over the other launching point, which he had similarly identified a few miles along the coast at Zempin. The photographs showed the two launching sites, and it happened that at Peenemunde we got a V1 on the ramp. Peenemunde was duly bombed in August 1943 and the damage was so great that production of the V1 was put back by six months.

On a more somber note, we mention that the inverse trajectory problem (based on the ocean floor debris scatter pattern) was an important element in the investigation of the crash of TWA Flight 800 in 1996.

1.4 Two Bodies

> In Newton's theory there is nothing about an inverse-square law that is particularly compelling.
>
> Steven Weinberg

Our word "revolution," used in the sense of a sudden overthrow of the established order, comes from the publication in 1543 of a book on astronomy by the Polish cleric Nicolaus Copernicus (1473–1543). *De Revolutionibus Orbium Coelestium* reasserted the heliocentric model of the solar system (first proposed by Aristarchus (310 B.C.–230 B.C.)) and argued that the earth revolves daily on its axis. Copernicus' theory was still circle-based and a second intellectual revolt was necessary to break the tyranny of the circle. It was the dogged determination of Johannes Kepler (1571–1630) to fit a curve to Tycho Brahe's (1546–1601) data on Mars that led him to his "physical" theory of the solar system.

Kepler, showing an impressive respect for Brahe's observational skill, let the data do the talking. And the data told him that the planetary orbits satisfy three laws:

1) The planets orbit the sun in ellipses, with the sun at one focus.
2) The radial segment from the sun to a planet sweeps out equal areas in equal times.
3) The square of the period of a planet about the sun is proportional to the cube of the length of the major axis of its orbit.

Kepler's laws are purely empirical; they do not suggest a physical reason for the observed oblong orbits. At the time, the prevailing belief was that planetary motions must result from contact forces that swept the planets around in their orbits, as in Descartes' theory of ethereal vortices. Newton's revolutionary idea was to abandon the need of a physical cause ("I frame no hypotheses" was his claim) and seek instead the mathematical form of the force law implied by the physical evidence. He came to his law of gravitation while sitting out the plague at his mother's Lincolnshire home in the years 1665–66. He said of the time:

> I decided that the forces which keep the planets in their orbs must be reciprocally as the squares of their distances from the centres about which they revolve.

Actually, the young Newton's derivation of the inverse-square law was based on a naive model of circular planetary orbits. Some years later the inverse-square law was "in the air." The law was apparently conjectured independently by Christian Huygens (1629–1695), Christopher Wren (1632–1723), and Robert Hooke (1635–1703) (a circumstance that led to an unpleasant priority dispute), but none could provide a demonstration of the inverse-square law based on Kepler's laws for orbits. A proper derivation of the inverse-square law, now called the inverse problem for orbits, did not come until 1679. This derivation led to Newton's universal theory of gravitation and his theory of motion as laid out majestically in his *Principia*, the most influential science book of all time.

Newton's law of motion, that the force on a body is the product of its mass and acceleration, is perhaps the best known physical law. It is the basis for the **direct** problem of dynamics: given the force on an object (the model) and appropriate initial conditions (the cause), find the subsequent motion (the effect). In the years prior to the publication of the *Principia*, Newton solved a number of **inverse** problems for orbits, that is, given the form of the orbit, he derived the form of the force law that would generate the given orbit. Newton organized his results in the *Principia*, where he solves many direct and inverse problems for orbits. In particular, he solves the *inverse Kepler problem*: only the inverse-square law can account for elliptical focus-directed motion and hence only an inverse square gravity law can explain Kepler's orbits.

1.5 Another World

> Would it not be admirable to ascertain the existence of
> a body which we cannot even observe?
>
> Jean Valz

By the early part of the nineteenth century, the clockwork universe of Kepler and Newton had become an article of faith. The planets moved in Keplerian ellipses and their regular motions were predictable by Newton's laws of motion and universal gravitation. But the clock seemed to be missing a few ticks. The outermost of the seven planets known at the time, laggardly Uranus, seemed determined to break Newton's laws. The sparse observational data on Uranus available at the time consisted of old observations by John Flamsteed (1646–1719) and others taken nearly a century before it was realized (in 1781) that the object observed was in fact a new planet, and of contemporary observations of Uranus published by Alexis Bouvard (1767–1843) in 1821. But the old observations did not square with Bouvard's new numbers. As Bouvard put it:

> ... if we combine the ancient observations with the modern, the former will be sufficiently well represented, but the latter will not be so, with all the precision which their superior accuracy demands; on the other hand, if we reject the ancient observations altogether, and retain only the modern, the resulting tables will faithfully conform to the modern observations, but will very inadequately represent the more ancient.... I leave to the future the task of deciphering whether the difficulty of reconciling the two systems is connected with the ancient observations, or *whether it depends on some foreign and unperceived cause which may have been acting on the planet* [my italics].

Bouvard's challenge was taken up by the brilliant young Englishman, John Couch Adams (1819–1892). During the summer vacation of his second undergraduate year at Cambridge, Adams set himself the task of solving the puzzle of Uranus. In a memorandum discovered among his papers after his death was found:

> 1841, July 3. Formed a design in the beginning of the week, of investigating, as soon as possible after taking my degree, the irregularities in the motion of Uranus, which are yet unaccounted for; in order to find whether they may be attributed to the action of an undiscovered planet beyond it; and if possible thence to determine the elements of its orbit, etc. approximately, which would probably lead to its discovery.

Some undergraduate!

The **direct** problem of finding the path of a planet given the masses influencing it was relatively well understood in Adams' time. What the young Cambridge undergraduate planned in 1841 was an attack on the much more difficult **inverse** problem: given the perturbations in the orbit of Uranus, determine the nature and location of an as yet undiscovered planet that would account for the perturbations. He not only planned to find the *cause* of the observed *effects* assuming Newton's laws as his *model*, he audaciously proposed that the results of his calculations could be used to actually *discover* the unknown planet!

Adams completed his solution of the inverse problem in 1845, but to prove his theory he would have to convince an observational astronomer to search that part of the heavens in which his calculations predicted the new planet would be found. To accomplish this, Adams would need help in high places. Adams petitioned the Astronomer Royal, George Biddell Airy (1801–1892), a naturally conservative man who was not given to changing his work schedule to set off on a possible wild goose chase suggested by the mathematical scribblings of a young Cambridge Fellow. Furthermore, Airy believed that the discrepancies in the orbit of Uranus might be explained by a different type of inverse problem, namely that of *model identification*. Perhaps the curious orbit of Uranus could be put down to a fault in Newton's model of gravitation, rather than the existence of a new planet. As Airy put it, "On this question therefore turned the continuance or fall of the theory of gravitation."

Adams' relationship with Airy was a litany of misunderstandings and missed opportunities. These misfortunes prevented Adams from being recognized as the discoverer of Neptune. For independently of Adams, the brilliant Frenchman Urbain LeVerrier (1811–1877) also solved the inverse problem of perturbations, arriving at a solution that differed only trivially from that of Adams. LeVerrier's coordinates were provided to the astronomer Johann Galle (1812–1910) in Berlin, who used them to make an undisputed sighting of the new planet on September 23, 1846.

An acrimonious priority dispute, reminiscent of the Newton–Leibniz brawl, broke out between English and French astronomers over who, Adams or LeVerrier, should be recognized as the discoverer of Neptune. In the long run the important point is that the solution of an inverse problem led to an historic scientific discovery. Moreover, the discovery and its mathematical explanation left little doubt that Newton's law of gravitation extended out to the far reaches of the solar system. Gravitation seemed indeed to be universal.

1.6 The Fountains of Dijon

> Like a person, the technique of groundwater modeling has two legs: the forward solution and the inverse solution. If one leg is long and another is short, how can the technique walk well?
>
> N.-Z. Sun

In the mid-nineteenth century the city fathers of Dijon commissioned a major improvement and enlargement of the town's public water works. The responsibility for carrying out this civil engineering program fell to Henry Darcy (1803–1858), the Inspector General of Bridges and Roads. Darcy's official report on the project, *Les Fontaines Publiques de la Ville de Dijon* (1856), was a monumental 647-page volume with an accompanying atlas of illustrations. One of the many practical problems faced by Darcy was the design of sand-filled filters for the public fountains. In particular, Darcy needed to design filters of sufficient capacity to meet the requirements of the growing city.

At the time there was no theoretical model for the flow of a fluid through a porous medium. Darcy was therefore compelled to personally conduct experiments on the flow rate of water through sand. His research on the topic was published as an appendix to his report titled "Determination of the law of flow of water through sand." Darcy's law, as first stated in the appendix, holds that the flow rate of water through a horizontal, homogeneous sand filter is inversely proportional to the length of the filter and directly proportional to the difference of the hydraulic heads at the ends of the filter, the constant of proportionality being dependent upon the particular type of sand in the filter.

Given the material parameter (the *transmissivity*) for the sand and the length of the filter, the **direct** problem of calculating the flow rate from the hydraulic head data is straightforward, and similarly the determination of the transmissivity for a single sand in a filter of given length from head and flow information is simple. The **inverse** problem of aquifer transmissivity, that is, the determination of a spacially varying transmissivity from flow and head data at test wells, is a hugely important problem. In this model identification problem the spatial inhomogeneity of the transmissivity results from the differing composition or porosity of the subterranean rock, sand, and soil. Unlike Darcy's uniform linear filter, the subterranean region of interest is three-dimensional. The desire to solve this inverse problem is driven by the need to solve the direct problem of groundwater flow. Without a properly identified model, the data necessary for solving the direct problem is simply unavailable. For example, only an accurate model of the inaccessible transmissivity field will allow the tracking of pollutants, a large-scale environmental problem of immense significance. Effective modeling of underground aquifers involves a delicate interplay between direct and inverse solution methods leading to successively more realistic representations of the subsurface.

1.7 The Universe

That which is far off, and exceeding deep, who can find it out?
Ecclesiastes 7:24

Prior to the eighteenth century, astronomers were primarily occupied with the solar system. In classical astronomy the stars were considered to be *fixed* to the "celestial sphere" and while the sphere executed a seasonal rotation with respect to the earth, the stars themselves were not believed to have "proper motions." The most conspicuous objects in the sky were the moon and the planets which wandered about on the celestial sphere.

By Edmond Halley's time the telescope had become a powerful tool for investigations of the heavens beyond the solar system. The new observations did not square well with the classical picture. First, it was clear that the stars have varying brightness, suggesting a distribution of the stars in space, rather than their confinement to a spherical surface. Also, Halley, in comparing his own measurements of declinations of stars with classical measurements recorded by Ptolemy, discovered that the stars have their own proper motions. The heavens then came to be recognized not as a static and perfect universe, but as a dynamic and heterogeneous structure. Furthermore, Halley's friend Christopher Wren was one of the first to speculate on the nature of the faint patches of light known as nebulae, suggesting that close observation might reveal "every nebulous star appearing as if it were the firmament of *some other world*, at an incomprehensible distance." This idea of extragalactic stellar systems as "island universes" was later taken up by the philosopher Immanuel Kant (1724–1804) and verified in observations by the astronomer William Herschel (1738–1822). In the twentieth century Edwin Hubble's (1889–1953) discovery of the red shift made the measurement of radial velocities of stars possible.

In the early part of the twentieth century two inverse problems related to stellar distribution and dynamics presented themselves. The first concerned the distribution of stars in globular clusters. These are spherical nebulae, which when observed from Earth appear as discs, namely the projection of the spherical cluster onto the observational plane (i.e., the photographic plate) of a telescope. Now the **direct** problem of determining the distribution of stars on the observational plane, given their three-dimensional distribution in space, is fairly straightforward. But of far greater interest to astronomers was the **inverse** problem of reconstructing the three-dimensional distribution that would account for the two-dimensional observations (shades of Plato's cave!). The other inverse problem concerned the distribution of proper motions of stars. Only *radial* velocities are observable from the earth, using spectral shift data, but of more interest is the distribution of proper, i.e., three-dimensional, velocities. The **direct** problem of determining the radial velocity distribution from knowledge of the proper velocity distribution is again fairly routine. The **inverse** problem of determining the proper velocity distribution from knowledge of the radial velocity distribution was solved in 1936 by the astronomer Viktor Ambartsumyan (1910–1996) when he rediscovered and applied a technique for reconstructing a function from its line integrals that was developed by Johann Radon in 1917 but was overlooked by the rest of the scientific community for nearly fifty years.

Because of the vast distances and hostile environments involved, direct on-site measurements are precluded in interstellar astronomy. The field will therefore continue to be an important area for the application of inverse theory.

1.8 Got the Time?

> . . . 'tis no where revealed in Scripture how long the earth had existed before this last Creation.
>
> Edmond Halley

In elementary differential equations courses a lot of time is spent on initial value problems. These are **direct** problems in which an initial value of a function is given along with a differential equation that models the evolution of the function. The goal is to find values of the function at later times. An interesting **inverse** problem involves *time reversal*: Find initial information from knowledge of the model and given states of the function. A favorite problem of this type is an old chestnut of Ralph Palmer Agnew:

> One day it started to snow at a heavy and steady rate. A snowplow started out at noon, going 2 miles in the first hour and 1 mile the second hour. What time did it start snowing?

Inverse chronology problems of this type require turning back the hands of a mathematical clock to find out when an event of interest occurred. Few, if any, problems of this type have received more attention than the geochronology problem. Of course, by "clock" we mean here a physical process whose behavior over time can be modeled mathematically.

Edmond Halley was one of the first to attack the inverse chronology problem for the age of the earth. The clock that Halley chose was the salinity of the sea. Halley's assumption was that when water originally condensed to form the oceans it was pure but that the action of rain and rivers to dissolve and transport minerals gradually increased the salinity of the seas over the eons. He reasoned that if measurements over time show that salinity increases, then "we may by Rule of Proportion, take an estimate of the whole time wherein the Water would acquire the Degree of Saltness we at present find in it." Halley proposed that the Royal Society commission salinity studies and carefully archive the results in a long-time salinity data base, and he lamented the fact that the ancient Greeks and Romans had left behind no salinity information. Halley's idea was first taken seriously by John Joly (1857–1933), who in 1899 made a study of sodium disposition in the seas. Making corrections (which turned out to be too conservative) for feedback mechanisms, such as recycling of salt from marine sedimentary rock and windborne salt blown inland, Joly arrived at a figure of about 90 million years for the age of the oceans, a figure now known to be very low. In fact, uncertainties of the volumes and compositions of ancient seas, continents, and atmospheres, and the details of the transport processes, make it unlikely that a sufficiently accurate salinity model can be constructed to resolve the inverse chronology problem for the oceans.

Another notable, but ultimately failed, attempt to apply inverse theory to geochronology was based on heat dissipation. The great physicist Lord Kelvin (William Thomson (1824–1907)), unsatisfied with what he believed to be the vague chronology espoused by the uniformitarianist geologists of his day, set for himself the program of establishing a geochronology on the firm mathematical base of Fourier's theory of heat. Jean-Baptiste Fourier (1768–1830) had developed a mathematical model for the flow of heat in a body in terms of a partial differential equation describing the evolution of the temperature of a body over time. Kelvin reasoned that Fourier's model could therefore be used as a natural clock: if one could accurately estimate current temperature information, and a parameter (the *conductivity* of the earth), then the clock could be turned back to find the time of a prior state, say when the earth was a molten ball. It had long been observed that temperature increased with depth in mines and Kelvin took as his starting point an estimate of the current temperature gradient at the surface of the earth, based on this temperature-depth data. He then worked backward in time to the point at which the earth was a molten ball at temperature 7,000°F. This led him to an estimate of less than 100 million years for the age of the earth. Kelvin's estimate is an historically interesting use of inverse theory in mathematical physics, but it turned out to be significantly on the low side because his model did not take account of some very important factors. In particular, Kelvin solved what we now call a *homogeneous* differential equation, but a very crucial inhomogeneity, namely a *source term* modeling the generation of heat from the decay of radioactive elements (unknown in Kelvin's day) was not included in his model.

As a final example of inverse chronology we briefly mention the radiocarbon dating technique for which Willard F. Libby (1908–1980) was awarded the 1960 Nobel Prize in chemistry. Unlike the unsuccessful attempts to solve the geochronology problem based on salinity or heat dissipation, Libby's method of dating artifacts took place on a laboratory scale and involved physical parameters that could be accurately estimated. We will not discuss the details of Libby's carbon-14 method as they can be found in many textbooks. Ultimately radiometric techniques, using long-lived nucleides, such as uranium-238, led to the currently accepted estimate of about 4.5 billion years for the age of the earth.

1.9 The Underworld

Now let us descend here below into the blind world.
Dante Alighieri

Eratosthenes could not measure the circumference of the earth directly, but by a brilliant use of indirect reasoning he solved the inverse problem of determining the angle from the arc and arrived at an impressively accurate estimate of the earth's girth. Over the centuries many investigators used indirect methods to refine the estimate of the earth's circumference. In 1669 Jean Picard (1620–1682), used stellar observations to determine the circumference within an error of less than .5 percent, so that in Newton's time an important parameter, the size of the earth, had been nailed down.

When we know the size of a thing our thoughts usually turn to its weight. A direct approach to this problem for the earth would involve taking many samples of the earth's "stuff" to determine an average density and hence a total mass. But clearly this approach was not suitable. For one thing, the sampling would be limited to the surface or near surface, and it was not at all clear in those days that the earth was not hollow. Newton's theory of gravitation suggested the possibility of an indirect approach to finding the total mass of the earth. Perhaps the mass could be deduced indirectly by carefully observing the relative gravitational effect of a large known mass on another mass, like a pendulum bob, suspended in the earth's gravitational field? In a footnote to the third book of the *Principia*, Newton mused about the possibility of a mountain causing a measurable deflection of a pendulum from the true vertical, but he opined that the effect would be unobservable. But if the deflection could be measured and the mass of the mountain estimated, then it would be possible to work out the mass of the earth using Newton's theory of forces.

A first attempt to estimate the mass of the earth based on Newtonian principles was made by Pierre Bouguer (1698–1758) in a series of experiments conducted in Peru in 1738. His experiments were inconclusive with respect to determining a numerical value of the earth's mass, but they did establish that a massive mountain apparently does deflect a pendulum bob from the vertical. This effect was also noticed by Charles Mason and Jeremiah Dixon when they conducted their historic survey of North America in 1766–68. In 1771, Nevil Maskelyne (1732–1811), the Astronomer Royal, proposed that the Royal Society mount an expedition to measure the "attraction of mountains" effect precisely. A saddle-back mountain in Perthshire known as Schiehallion was chosen as the site of the experiment. During four arduous months in 1774 Maskelyne and his assistant Reuben Barrow measured the deflection of the pendulum and prepared accurate contour plots to estimate the volume of Schiehallion. Using Maskelyne's data and assuming that the average density of the mountain was the same as common stone (about 2.5 times that of water), Charles Hutton, Professor of Mathematics at the Royal Military Academy, worked out that the mean density of the earth is about 4.5 times that of water. The average density of the earth was therefore nearly twice that of common stone and hence the underworld must be very dense indeed. This put to an end the lingering belief, going back to ancient legends of Hades, that the interior of the earth is hollow.

In 1798 Henry Cavendish (1731–1810) conducted a laboratory experiment that refined the estimate of the earth's mass. The experiment involved measuring indirectly the gravitational attraction of lead balls in a finely balanced torsional pendulum. With his apparatus, Cavendish arrived at an estimate for the universal gravitational constant and from this he was able to set the mean density of the earth at about 5.4 g/cm^3. Today the accepted density is about 5.5 g/cm^3 giving a total mass for the earth of about 5.97×10^{24} kg.

In the ninteenth century the invention of the seismograph gave scientists a new tool for indirectly probing the earth's interior by using dynamic earthquake data. Although it wasn't possible (before the invention of nuclear charges) to thump the earth as you would a melon, it was possible to observe with a seismograph the effects of such thumps which occurred in the form of earthquakes. Imagine an earthquake occurring near the surface at some point on the globe. Seismic waves would fan out from the quake traveling through the earth's interior and would be measured by seismographs in observatories at various points on the earth's surface. In a paper titled "The Constitution of the Interior of the Earth, as Revealed by Earthquakes," R. D. Oldham (1858–1936) noticed a startling fact in the seismological records and indirectly deduced a surprising result. If the center of the earth is taken as the origin of an xy-coordinate system and the quake is placed at the point $(1,0)$, then Oldham noticed that in a "shadow region" between polar angles 105° and 142° seismological stations do not detect the quake. Treating seismic waves as light rays Oldham realized that materials of different elastic properties refracted the waves differently and by backtracking rays from the boundary of the shadow region he deduced that the earth has a distinctive central core which bent the seismic waves and formed the shadow region. This was the first application of inverse techniques to lead to a structure theory for the interior of the earth.

Since Oldham's time, inverse methods have continued to yield clues about the earth's interior. The earth's core is now known to consist of two major parts, a fluid outer core of diameter about 4,350 miles and a solid inner core of about 1,500 miles diameter. Analysis of decades of seismic data has revealed that seismic waves move through the earth faster on a north–south course than they do on east–west courses and a roughly north–south "fast-track axis" has been identified. Quite recently it was discovered that this fast-track axis rotates around the earth's pole indicating that the inner core itself rotates at a *faster* rate than the earth. Geophysicists hope that this new knowledge of the dynamics of the inner core will lead to a better understanding of the earth's magnetic field, and in particular, might help clear up the mystery of the periodic reversal of the magnetic poles.

1.10 Head Games

> ... this seemed like a problem which would have been solved before, probably in the 19th century...
>
> A.M. Cormack

Given a real-valued function defined on a plane, we can define a function on all lines in the plane by computing the line integral of the fixed function over a given line. This is a standard **direct** problem: a function of three variables (e.g., the slope and the coordinates of a point defining a line) is produced by integrating a function of two variables. This transformation from functions on a plane to functions on all lines in the plane is called the Radon transform, after the Austrian mathematician Johann Radon (1887–1956), who solved the **inverse** problem. That is, Radon devised a formula for reconstructing a function given its line integral over all lines in the plane. Radon's publication, in 1917, of the solution of this purely mathematical inverse problem was little noticed and his inversion formula was rediscovered several times in later years. His inversion formula turned out to have immense practical importance and it eventually led to the awarding of a Nobel Prize in Medicine to Allan Cormack and G. N. Hounsfield in 1979. Here's the story:

A century ago medical diagnostic options were very limited. If a patient was suspected of having a tumor, then exploratory surgery was the normal course, and such surgery could often be a stab in the dark. (Forgive me.) Clearly more informed *noninvasive* diagnostic procedures would be preferred, particularly when dealing with sensitive areas such as the brain. One such procedure became available when Wilhelm Roentgen (1845–1923) developed X-ray technology. In a conventional X-ray the beams emerge from a fixed source, pass through the body and strike a photographic film. On the resulting picture many details are hidden; the images of bones, organs, tumors, etc., are projected onto a two-dimensional plane and they overlap. The situation is highly analogous to that discussed by Plato nearly 2500 years ago.

In 1955 Allan Cormack, a young lecturer at the University of Cape Town, was asked to fill in part-time as staff physicist at Groote Schuur Hospital (later to become famous as the site of the first human heart transplant). Part of Cormack's job was to prepare isodose charts for radiotherapy. These charts were constructed by trial and error methods assuming the body was homogeneous. But the body is anything but homogeneous and, as Cormack tells it, "It occurred to me that in order to improve treatment planning one had to know the distribution of the attenuation coefficient of tissues in the body, and that this distribution had to be found by measurements external to the body. . . ." Furthermore, Cormack realized, ". . . this information would [also] be useful for diagnostic purposes . . . ."

As an X-ray beam passes through tissue and bone, its interaction with the intervening material causes the ray to be attenuated—weakened. The degree of attenuation depends on the nature, density, and extent of the material that the beam encounters. This can be explained in terms of a function that gives for each point of the body a number, the *attenuation coefficient*, which is a measure of the rate at which the beam is attenuated at that point. The physical nature of the body gives rise to this attenuation coefficient function and hence a plot of this function provides a "map" of the interior of the body. The total attenuation of a beam in a given direction can be measured and related to the line integral of the attenuation coefficient function over the path of the beam. To construct the map of the interior, that is, the CAT scan, one must recover the attenuation function from knowledge of its line integral over all lines. This is just Radon's problem and its application in medical technology is called *computed tomography*. The difference between computed tomography and conventional X-ray methodology is that in CAT, instead of obtaining an image from a single aspect, projections are taken from many different angles and the results are mathematically "back projected" to reconstruct the object—a classic inverse technique.

Since the work of Cormack and Hounsfield in the sixties and seventies the field of medical imaging has made great advances. Many new indirect, noninvasive diagnostic procedures have been developed. Foremost among these may be the magnetic resonance imaging (MRI) technique. In this method the imaging of the interior of the body is accomplished by the use of magnetic fields rather than with potentially harmful ionizing radiation. In MRI, proton density, rather than the attenuation function, is used as the medium to map the interior of the body. Roughly speaking, the protons may be considered to be tiny magnets that are distributed, with varying density, throughout the body. In the MR imager a strong, stationary magnetic field aligns these little magnets along the axis of the scanner and a microwave pulse sets them rotating. Faraday's law relating a varying magnetic field with the induced current is the basis of the method and the image is constructed by solving the inverse problem of reconstructing the proton density from the observed induced currents in the pickup coil of the MR imager.

1.11 Why Teach Inverse Problems?

The vignettes above are meant to convince the reader that the solution of inverse problems has been an important factor in the development of mathematics and the sciences. This reason alone would justify the occasional study of inverse problems in the early years of the undergraduate curriculum. But more important than the study of any specific inverse problems is the inculcation of a habit of "inverse thinking" in students. If students consider only the direct problem, they are not looking at the problem from all sides and will fail to see two-thirds of the whole picture. The habit of always looking at problems from the direct point of view is intellectually limiting and the very *posing* of an occasional inverse problem is an affirmative act of questioning that should always be encouraged on the student's part.

Not all the modules in this guide are deep—some are in fact fairly superficial. But the hope is that the guide will provide the instructor with a collection of tools that cross curricular lines and can occasionally be used in many different elementary courses. Taken as a whole, these materials should be useful in kindling physical intuition, fostering model building, encouraging conjecturing, stimulating the use of basic mathematical analysis and computational techniques, and generally promoting critical thinking by students in the first two undergraduate years.

I will close this chapter with my favorite example of a nonmathematical inverse problem that illustrates the power of "inverse thinking." The story begins in the spring of 1947 when Mohammad adh-Dhib, a Bedouin shepherd boy, while searching for a lost goat, stumbled upon a number of earthenware jars containing ancient manuscripts in a cave at Qumran in the Judean desert near the Dead Sea. In a tale worthy of Indiana Jones, a "scroll rush" developed as various museums and jurisdictions vied with each other to collect as much scroll material as possible (four scrolls were offered for sale in a classified ad appearing in the June 1, 1954, edition of the *Wall Street Journal*). The bulk of the material now known as the Dead Sea Scrolls came under the control of an international committee set up by the Rockefeller Museum, the Ecole Biblique et Archéologique, and the Jordanian Department of Antiquities. The small group of scholars on the committee, who held lifetime sinecures, hoarded the scroll material and published their own findings at a glacial rate while denying other scholars access to the material. This, needless to say, was a source of considerable frustration for the rest of the scholarly community. The committee did, however, produce a concordance of the scroll material that was available to independent scholars. The production of the concordance was a **direct** problem: the *input* was the text of the scrolls themselves, the *process* for producing the concordance was well understood, and the *output* was the concordance. Four decades after the production of the concordance, Ben-Zion Wacholder and Martin Ebegg, of Cincinnati's Hebrew Union College, thought about the problem of scroll access inversely and conceived the idea of writing a computer program to reconstruct an estimate of the original scroll text using the concordance as data (I like to call this *literary tomography*). The Wacholder–Ebegg reconstruction was close enough to the original text that the committee was forced to concede and allow publication of the original material. The front page of the September 22, 1991 edition of *The New York Times* announced that "the monopoly over the Dead Sea Scrolls is ended." This was a brilliant triumph of inverse thinking.

1.12 Notes and Suggestions for Further Reading

Aristotle's argument for the sphericity of the earth appears in Chapter 14 of Book II of his work *On the Heavens*. A wonderfully readable account of Eratosthenes' measurement of the earth and extensions to other methods of mapping the earth and the universe can be found in Osserman (1995). Drake and Drabkin (1969) is a masterful history of early ballistics, including the work of Tartaglia (see also Drake (1970)). Galileo (1638) speaks best for himself on the birth of mechanics.

I feel that Halley is not generally appreciated for the great man that he was. This is perhaps natural; Newton casts a very big shadow. Armitage (1966) and Cook (1998) provide complete treatments of Halley's life and work. His gunnery rule is presented from a modern point of view in Groetsch (1997). The story of R. V. Jones' V1 trajectory backtracking is told in Wintherbotham (1974).

In looking through the *Principia*, it is hard not to get the idea that Newton was having fun with the inverse problem for orbits (centrally directed elliptical orbits, spiral orbits, etc.). What we now call the *inverse* problem for orbits was originally called the *direct* problem (see Brackenridge (1996))—times, and words, change. Like all of Gamow's books, his popular book on gravity (Gamow (1962)), which discusses Cavendish's experiment, is a delight to read. Rouse Ball (1893) is still one of the best books *about* the *Principia* and its origins, while Westfall's (1980) biography of Newton is hard to beat.

The Adams–LeVerrier competition is a fascinating scientific and human story well told by Grosser (1962) (see also Jones (1947) for another popular account). A bit more mathematics pertaining to the problem can be found in Bollobás (1986).

Hubbert (1969) covers both the history and the basic science of ground-water motion; it also contains a reprint of the appendix of Henry Darcy's report.

Although it appears in a biology journal, Wicksell (1925) seems to be the earliest source on the globular cluster problem. More on Hubble and the red shift can be found in Christianson (1995).

The age-of-the-earth problem was quite controversial in Victorian times because on it hinged the viability of Darwin's theory of evolution (which required a very old earth for it to work). An account of the controversy can be found in Burchfield (1975). Eicher (1968) is an excellent introductory presentation of geochronology and Holmes (1913) contains a lot of interesting history on the problem.

The story of Schiehallion and the deflection of the pendulum is told in Howse (1989). Poynting (1913) is a dated, but still very useful, reference on the basic physics of the earth. Burger (1992) is a very good modern source on the fundamentals of remote sensing of the shallow subsurface of the earth. For introductory information on the deeper interior, Bolt (1982) is excellent. Broad (1996) is a news story on the still unfolding investigation of the deep interior.

Allan Cormack (1980, 1982) gives firsthand reports on the birth of computed tomography (Johann Radon's paper is reprinted in the same volume that contains Cormack (1982)). Gordon, Herman, and Johnson (1975) is a readable popular article on computed tomography, and Shepp and Kruskal (1978) treats the elements of the mathematical theory. Peterson (1986) is a brief popular article giving a different slant on tomography.

General introductions to inverse problems can be found in Groetsch (1993) and Kirsch (1996). These two works are at a more advanced mathematical level than this book.

2

Inverse Problems in Precalculus

Not all inverse problems involve advanced mathematics. Fascinating, and at times challenging, inverse problems can arise in quite simple mathematical models of basic physical processes. The modules of this chapter investigate some elementary inverse problems making use of simple algebra, analytic geometry, and trigonometry. The physical models involved include Galileo's law of falling bodies, Torricelli's law governing the velocity of an effluent, parabolic trajectories, Newton's law of gravitational attraction, and the relationship between speed, distance, and time. Although the mathematics connected with the activities of this chapter is elementary, some of it requires the use of a graphing calculator or even a calculator with a symbolic algebra processor.

2.1 A Little Squirt

Course Level:

Precalculus (Algebra)

Goals:

Explore existence and uniqueness of solutions.

Mathematical Background:

Parametric equations, quadratics

Scientific Background:

Law of falling bodies, Torricelli's Law

Technology:

Graphing calculator

2.1.1 Introduction

Fill a tank with water, then drill a hole in its side. What happens? Water squirts from the hole, follows a graceful arc, and splashes down some distance from the tank. Simple physical intuition tells us that the distance reached by the spurt (i.e., the range of the leading drop of the spurt) is clearly determined by two things: the height of the hole and the velocity with which the water emerges from the hole. The velocity is determined by the pressure at the level of the hole, which is in turn dependent upon the *hydraulic head*, that is, the height of the column of water above the hole. The relationship was discovered by Galileo's protégé Evangelista Torricelli (1608–1647).

Let us agree to neglect the effect of air resistance and assume that the only force acting on any of the drops of water that emerge from the hole is the constant force of gravity. Suppose that the tank is situated on a horizontal plane and that coordinate axes are chosen as in Figure 2.1.

Imagine a drop of water emerging from the hole with horizontal velocity v. As there is no force acting in the horizontal direction (remember, we are neglecting air resistance), the drop will always have velocity v and therefore in time t its horizontal position will be

$$x = vt.$$

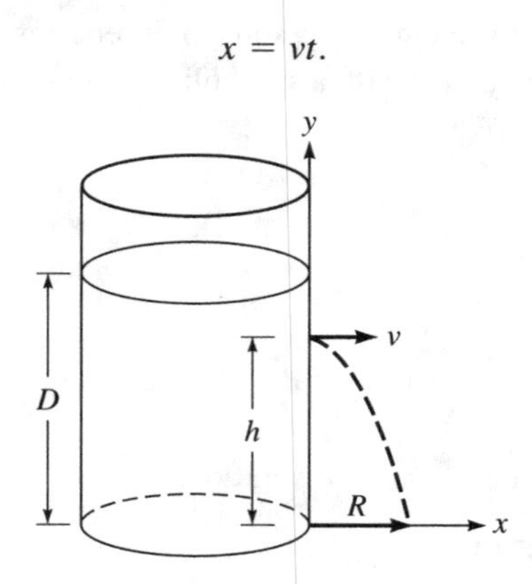

Figure 2.1: A Squirt

While the drop travels horizontally it continually falls under the influence of gravity and its vertical position is given by Galileo's law for bodies falling in a vacuum:

$$y = h - \frac{g}{2}t^2,$$

where g is the acceleration due to gravity. The horizontal velocity with which the water leaves the hole is given by Torricelli's Law:

$$v = \sqrt{2g(D - h)},$$

where D is the depth of the water in the tank and h is the height of the hole above the bottom of the tank. The quantity $D - h$, the depth of the column of water above the hole, is called the hydraulic head. Torricelli's Law is a consequence of the principle of conservation of energy. Imagine what happens during an instant when the hole is first unplugged. The water level goes down by a tiny amount, decreasing the volume of water in the tank by a tiny amount. An equal tiny volume of water is expelled from the hole. If the mass of this infinitesimal volume of water is m and its velocity is v, then its kinetic energy is $mv^2/2$. If frictional forces are ignored, then total energy is conserved and hence this kinetic energy must match the potential energy lost by the water in the tank. This potential energy is $mg(D - h)$, and hence

$$\frac{1}{2}mv^2 = mg(D - h),$$

which gives Torricelli's Law.

The direct problem for Torricelli's Law is to determine the range of the spurt R, given D and h. A simple inverse problem consists of determining the height h of the hole given the range R. This and related questions are investigated in the activities below.

2.1.2 Activities

1. Question The curve that the spurt follows can be expressed parametrically in terms of time. What is the shape of this curve?

2. Exercise Find the time of descent of the leading edge of the spurt (i.e., the time it takes for the leading edge of the spurt to hit the ground). Note that the time of descent is independent of D. Give a physical explanation of this independence.

3. Calculation Suppose the depth D of the water in the tank is 12 feet and the height h of the hole is 4.5 feet. Plot the trajectory of the spurt.

4. Problem Find the range R, that is, the x-intercept of the trajectory, in terms of D and h.

5. Question Given a depth D and a hole height h, would the resulting spurt attain (neglecting air resistance) a longer range on the earth or on the earth's moon?

6. Problem If the depth of water in the tank is D, find the interval of all possible ranges that may be attained for holes at various heights.

7. Calculation Suppose $D = 12$ ft. Plot the range R as a function of h for $0 \leq h \leq D$. Describe the shape of this graph—be specific.

8. Problem For a given R with $0 \leq R \leq D$, show that there are generally two hole heights h that produce the range R. Give a physical explanation of this.

9. Question Under what conditions does the inverse problem discussed in Problem 8 have a unique solution h?

10. Problem Suppose the tank has a single hole at height h and for two different water depths $D_1 > h$ and $D_2 > h$ the corresponding ranges R_1 and R_2 are measured. Show that D_1, D_2, R_1, and R_2 uniquely determine h.

11. Problem Show that for $0 < h < D$, the leading edge of the spurt does not strike the ground "vertically."

2.1.3 Notes and Further Reading

For more about Torricelli's Law as a demonstration in the classroom, see T. Farmer and F. Gass, "Physical demonstrations in the calculus classroom," *The College Mathematics Journal* **23** (1992), pp. 146–148 and R. D. Driver, "Torricelli's law: an ideal example of an elementary ode," *American Mathematical Monthly* **105** (1998), pp. 453–455. Other inverse problems related to Torricelli's Law may be found in C. W. Groetsch, "Inverse problems and Torricelli's law," *The College Mathematics Journal* **24** (1993), pp. 210–217. For biographical information on Torricelli, see P. Robinson, "Evangelista Torricelli," *Mathematical Gazette* **79** (1995), pp. 37–47.

2.2 A Cheap Shot

Course Level:

Precalculus (Algebra, Coordinate Geometry)

Goals:

Develop a simplified model of projectile motion. Study the "inverse range" problem. Investigate existence and uniqueness of solutions.

Mathematical Background:

Parabolas, hyperbolas, quadratics

Scientific Background:

Law of inertia, law of falling bodies

Technology:

Graphing calculator

2.2.1 Introduction

In the early sixteenth century, problems of gunnery were technical concerns of the first rank for Renaissance rulers. The new cannon had proved its deadly efficacy in the many interprincipality wars of the time. Furthermore, the cannon was looked to as the weapon that would provide one of the surest defenses against the most terrifying extra-European threat—the Ottoman Empire. Early studies of gunnery conducted in Renaissance Italy led to the modern theory of dynamics and hence may be viewed as lying at the origin of mathematical physics.

The first western book on gunnery, *New Science*, was published in Venice in 1537 by Nicolo Tartaglia (Nicolo Fontana, 1500?–1557). Tartaglia claimed to be the inventor of the gunner's square, a simple device for measuring the angle of elevation of a gun barrel with respect to the horizontal. The square allowed Tartaglia to study the *direct* problem of gunnery: given the angle of elevation, find the horizontal range (for a given powder charge). It also suggested the study of the corresponding *inverse* problem: Find the angle of elevation leading to a given range.

Tartaglia's analysis of projectile motion was flawed, but a century later Galileo provided a correct analysis of the motion of a point projectile in a vacuum. Galileo's achievement rested firmly on a foundation of three scientific pillars: his law of *inertia*, his law of *falling bodies*, and the idea of *composition* of motions. Suppose a point projectile of unit mass is launched from the origin at an angle θ to the positive x-axis with initial speed v. Galileo's idea was that the motion with speed v in the direction θ could be decomposed into two independent motions in the x- and y-directions with speeds v_x and v_y, respectively (see Figure 2.2).

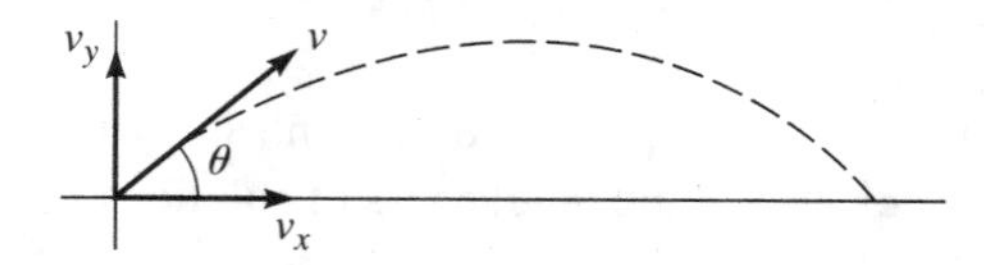

Figure 2.2: A Shot

The relationships between v and its components are the basic tenets of trigonometry:

$$\frac{v_x}{v} = \cos\theta, \quad \frac{v_y}{v} = \sin\theta.$$

Consider now the motion in the x-direction only. Since no force acts in this direction (remember, we neglect air resistance, and gravity acts only in the y-direction), Galileo's law of inertia holds that the velocity in the x-direction is constant. But the initial velocity in the x-direction is v_x $(= v\cos\theta)$, and hence the velocity in the x-direction is *always* v_x. Therefore, after t units of time, the projectile has moved to a position whose x-coordinate is

$$x = v_x t.$$

In the y-direction, two laws hold sway. The law of inertia would carry the projectile to a height $v_y t$ in t units of time. However, during all of this time the projectile is also *falling* in the earth's gravitational field. According to Galileo's law of falling bodies, during t units of time the body will *fall* $gt^2/2$ units of distance, where g is the constant acceleration due to gravity. Combining these effects we find that in t units of time the y-coordinate of the projectile is

$$y = -\frac{g}{2}t^2 + v_y t.$$

These two equations:

$$x = v_x t, \quad y = -\frac{g}{2}t^2 + v_y t,$$

give a *parametric* representation of the curve, or the *trajectory* that the projectile will follow.

2.2.2 Activities

1. Exercise Find the *time of flight* of the projectile in terms of v_y and g. (The time of flight is that positive value of t for which $y = 0$; in other words, it is that value of the time parameter corresponding to the positive x-intercept of the trajectory.)

2. Exercise The *range* R of the projectile is the positive x-intercept of the trajectory. Show that the range satisfies $v_x v_y = gR/2$.

3. Problem Show that any nonnegative range is, in principle, achievable for infinitely many combinations of the x- and y-components of the initial velocity, v_x and v_y.

4. Calculation Take $g = 32.2 \text{ ft/sec}^2$ and suppose the range R is 5,000 feet. Plot the set of points $\{(v_x, v_y)\}$ for which the range is attained. What is the shape of this "initial velocity curve?" Find the horizontal component of the initial velocity v_x for which the range $R = 5{,}000$ is achieved, given that $v_y = 400$ ft/sec. Given that $v_x = 300$ ft/sec, find the vertical component of velocity v_y for which the range $R = 5{,}000$ is achieved. Use the trace feature on your calculator to estimate the point on the "initial velocity curve" for the range $R = 5{,}000$ that is nearest to the origin (be sure that your display is "square"). This is the point on the curve for which the desired range is achieved and the initial kinetic energy $(v_x^2 + v_y^2)/2$ is smallest. This particular aiming would therefore achieve the desired range with the smallest expenditure of powder charge. Estimate the angle of elevation θ that corresponds to this minimal energy shot.

5. Problem Show that for a given muzzle velocity v, the range R is given by

$$R(\theta) = \frac{v^2}{g} \sin 2\theta.$$

6. Question For a given muzzle velocity, what angle of elevation produces the maximum range? What is the maximum range?

7. Problem Suppose the muzzle velocity v is fixed. Use "completion of the square" to find the maximum altitude of the projectile in terms of the angle of elevation and muzzle velocity.

8. Problem Suppose the range $R > 0$ is specified. What angle of elevation will lead to this range for a minimum value of the muzzle velocity v? What is the minimum muzzle velocity (in terms of R)? (Compare with Calculation 4.)

9. Problem Suppose the muzzle velocity v is fixed. Show that each range that is smaller than the maximum range is attainable with exactly two distinct angles of elevation. How are these angles related?

Consider now an uphill battle where the point projectile of unit mass is fired from the origin with muzzle velocity v at an angle θ to the horizontal and strikes a plane that is inclined at an angle $\alpha < \theta$ to the horizontal, as shown in Figure 2.3.

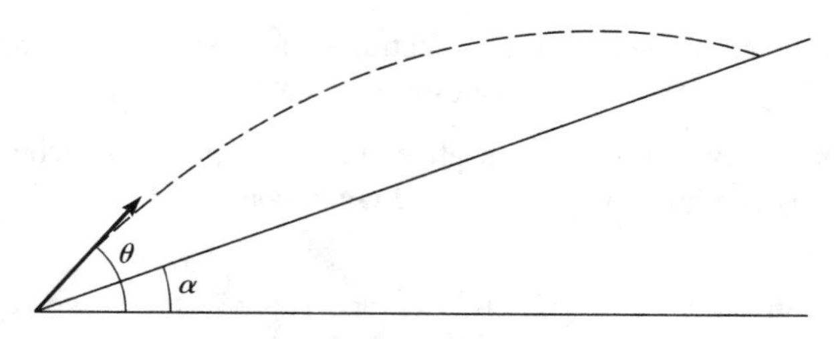

Figure 2.3: An Uphill Battle

10. Problem For a given muzzle velocity v and angle of elevation θ, find the time of flight to the target on the inclined plane. Compare with Exercise 1.

11. Problem For a given muzzle velocity v and battlefield angle α, find the horizontal range R (i.e., the positive x-coordinate of the intersection of the trajectory with the inclined plane) as a function of the horizontal and vertical components of the muzzle velocity. Compare with Exercise 2.

12. Calculation Suppose $g = 32.2\ \text{ft}/\text{sec}^2$ and $\alpha = \pi/3$ radians. Plot the initial velocity curve $\{(v_x, v_y)\}$ of all pairs of muzzle velocity components that lead to a horizontal range of $R = 1,000$ ft. Estimate the point on this curve that is nearest to the origin (this corresponds to the minimally charged shot that attains the required range). To what angle of elevation θ does this minimal energy shot correspond? Repeat all of the above for a battlefield angle of $\alpha = \pi/6$. Do you observe a connection between the angle of elevation corresponding to the minimal-energy shot and the battlefield angle?

13. Problem Express the horizontal range R as a function of the muzzle velocity v, the gravitational acceleration g, the angle of elevation θ, and the battlefield angle α. Compare with Problem 5.

14. Problem For a given muzzle velocity v and battlefield angle α, find the angle of elevation θ leading to a maximum horizontal range R. Give a geometrical interpretation of your result. Compare with Question 6. (Hint: Pull out your table of trig identities!)

15. Problem Show that each nonmaximal range is attained by exactly two distinct angles of elevation. How are these angles related? Compare with Problem 9.

16. Project Suppose projectiles are fired from a fortification of height h, which sits atop a sloping battlefield as in Figure 2.4. Carry through as much of the development of the activities above as you can for this situation.

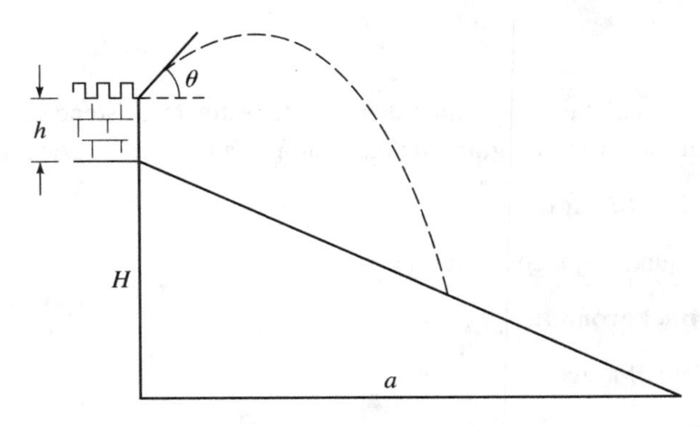

Figure 2.4: A Defensive Battle

2.2.3 Notes and Further Reading

Tartaglia considered the result of Problem 9 to be one of his greatest discoveries. In the letter of dedication of his *New Science* he says

> I knew that a cannon could strike in the same place with two different elevations or aimings. I found a way to bring this about, a thing not heard of and not thought by any other, ancient or modern.

References and more background on Tartaglia's work can be found in the introduction to C. W. Groetsch's "Tartaglia's inverse problem in a resistive medium," *American Mathematical Monthly* **103** (1996), pp. 546–551. The uphill battle in Activities 10–15 was studied by Edmond Halley of comet fame (see C. W. Groetsch, "Halley's gunnery rule," *The College Mathematics Journal* **28** (1997), pp. 47–51). Problem 15 greeted the bachelor candidates of Cambridge University when they sat on Friday morning, June 5, 1908, for the famous *Mathematical Tripos* exam. A vivid illustration of an ancient Hittite fortification at Boghaz Köy in Turkey that has the characteristics of Project 16 can be seen in Michael Grant's *In Search of the Trojan War*, Facts on File Publications, New York, 1985, pp. 170–171.

2.3 das Rheingold

Course Level:

Precalculus (Algebra, Coordinate Geometry)

Goals:

Integrate physical and graphical reasoning. Investigate existence and uniqueness of solutions. Employ graphing calculator to analyze a physical problem.

Mathematical Background:

Quadratic equations, inequalities, parabolas

Scientific Background:

Newton's law of gravity

Technology:

Graphing calculator

2.3.1 Introduction

This module is inspired by the ancient Germanic myth of the Niebelungen, a race of diminutive malefactors whose golden hoard was stashed at the bottom of the Rhine river. We treat the problem of locating and identifying a single isolated gravitational point source. Suppose a point mass m, say a nugget of gold, lies at the bottom of a calm river that is 1 unit deep. We make the simplifying assumption that all other gravitational sources are purely homogeneous so that the gravitational anomaly generated by the point source will be considered to be the only true gravitational effect. The determination of the gravitational force on a unit mass on the surface of the river engendered by the nugget at the bottom of the river is a very simple direct problem. Newton's law of gravitation holds that this force is equal to a known constant (the gravitational constant) times the product of the masses, divided by the square of the distance separating the masses. This is the famous "inverse square" law of gravitational attraction.

In this module we consider the inverse problem of determining the mass and location of a single nugget from measurements taken at the surface. The measurements consist of a distance x from a reference point on the surface and an estimate μ (obtained, say, by use of a delicate spring scale) of the vertical component of the gravitational force on the unit mass at position x on the surface engendered by the nugget below the surface. The situation is illustrated in Figure 2.5.

The square of the distance between the source nugget and the unit mass on the measuring device is given by the Pythagorean theorem, $1 + (x - s)^2$, and the product of the masses is $1m$, where m is the mass of the nugget. The vertical component, μ, of the gravitational effect at the position x on the surface

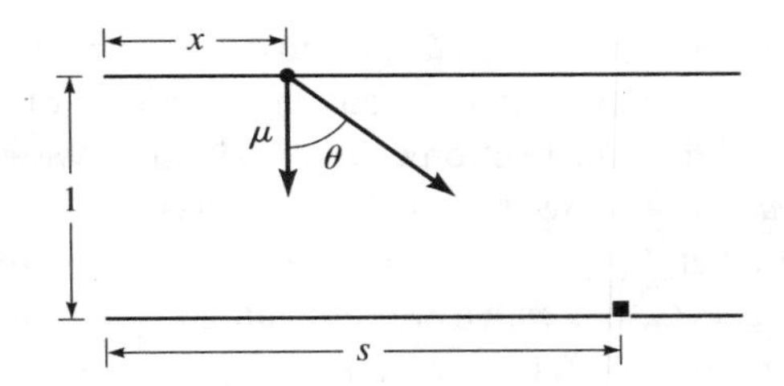

Figure 2.5: Vertical Force Engendered by a Point Source

is therefore, by Newton's law of gravitation,

$$\mu = \gamma \frac{m}{1 + (x - s)^2} \cos \theta,$$

where γ is the gravitational constant and θ is the angle pictured. Since the depth of the river is 1, we see from the figure that

$$\cos \theta = \frac{1}{\sqrt{1 + (x - s)^2}},$$

and substituting this above we get

$$\mu = \gamma m (1 + (x - s)^2)^{-3/2}.$$

The *direct* problem in this context is to determine the vertical force μ on the unit mass at position x on the surface that the mass m at position s at the bottom of the river engenders. This direct problem clearly has a unique solution given by the equation above. We will consider the *inverse* problem of determining the mass m and position s of the source from observations of the force μ at surface sites x. Before getting into this problem we will reformulate the equation to make our problem a bit simpler. Define new variables M and G, which we will call the *effective mass* and *effective vertical force*, respectively, by

$$M = m^{2/3} \quad G = \left(\frac{\mu}{\gamma}\right)^{2/3}$$

A measurement of μ then uniquely determines G and knowledge of M uniquely determines m. With these definitions, the equation above is easily seen to be equivalent to

$$M - G = G(x - s)^2.$$

The inverse problem now is equivalent to determining M, from which the mass m of the nugget can be obtained, and the location s of the nugget from knowledge of x and the effective force G at position x. We will call the pair (x, G) an *observation* because it consists of observing the effective force G (obtainable from μ) at the site x. The inverse problem therefore is equivalent to determining a pair (s, M), which we will call a *source*, from observations (x, G). If a unique source (s, M) is determined, then we have found the location s and the mass $m = M^{3/2}$ of the inaccessible nugget without getting wet—a feat that would surely arouse the envy of the Rhine Maidens!

2.3.2 Activities

1. Question Does a single observation (x, G) uniquely determine the source (s, M)?

2. Question Suppose s is plotted on a horizontal axis and M is plotted on a vertical axis. What is the shape of the *source curve* associated with a given observation (x, G)? (In other words, what is the graph of all possible single-point sources (s, M) that could account for the observation (x, G)?)

3. Question How does the shape and position of the source curve change with changes in the observation (x, G)?

4. Calculation Plot the source curve associated with the observation $(1, 2)$.

5. Problem Suppose an observation (x, G) is given. (a) Show that for every number $M > G$ there are two sources with effective mass M that can account for the observation. (b) Show that there is a unique source of effective mass $M = G$ that can account for a given observation (x, G). What is the location of this source? (c) Show that if $M < G$, then no source of effective mass M can account for the given observation.

6. Exercise Explain Problem 5 in intuitive physical terms rather than in mathematical terms.

7. Question Suppose observations (x_1, G) and (x_2, G) are recorded at distinct sites $x_1 \neq x_2$. What is the location of the source?

8. Exercise Find all sources (s, M) that can account for both of the observations $(0, 1)$ and $(1/\sqrt{2}, 2)$.

9. Question Is $\{(0, 1), (2, 6)\}$ a possible pair of observations? In other words, is there a single point source (s, M) that can engender both of the observations $(0, 1)$ and $(2, 6)$?

10. Exercise Find all sources that can account for the pair of observations $\{(0,1),(1,2)\}$.

11. Calculation Plot the source curve for the observation (1.12, 2.7). On the same axes, plot the source curve for the observation (3.1, 4.89). Estimate the sources (position and effective mass) that give rise to these observations.

12. Calculation Estimate the point on the source curve engendered by the observation (2.1, 4) that is closest to the origin.

13. Calculation Estimate all sources that can generate the observations $(-1.1,\ 2.2)$ and (0.9, 8.9).

14. Calculation Are the observations (1.2, 3.4), $(-2.1,\ 1.1)$, and (3.07, 2.7) consistent? (In other words, is there a source (s, M) that generates these observations?)

15. Question What is the largest number of possible sources that can account for a set of two or more observations at distinct sites?

16. Problem Find conditions on distinct observations (x_1, G_1) and (x_2, G_2) for which (a) the observations are inconsistent, (b) the observations determine a unique source, or (c) the observations may be accounted for by two distinct sources.

17. Problem Suppose two observations with $G_1 \neq G_2$ *uniquely* determine a source (s, M). Show the following:

(a) The distance between the observation sites is

$$|x_1 - x_2| = \frac{|G_1 - G_2|}{\sqrt{G_1 G_2}}.$$

(b) The source is located at

$$s = \frac{G_1 x_1 - G_2 x_2}{G_1 - G_2}.$$

(c) The effective mass is

$$M = G_1 + G_2.$$

18. Problem Show that at most one source can be located between distinct observation sites (i.e., given observations (x_1, G_1), (x_2, G_2) with $x_1 \neq x_2$, there can be at most one source (s, M), with s between x_1 and x_2, that gives rise to the observations).

19. Exercise Find the source that gives rise to the observations $\{(3, 10), (5, 2), (6, 1)\}$.

20. Problem

(a) Given an observation (x_0, G_0), show that by an appropriate choice of coordinate system we may assume that $x_0 = 0$, and by a suitable choice of units we can arrange that $G_0 = 1$.

(b) Let (s_1, M_1), (s_2, M_2) be two points (with $s_1 \neq s_2$) on the source curve $M = 1 + s^2$ generated by the observation in part (a). Show that the only other source curve passing through the points (s_1, M_1) and (s_2, M_2) is the curve

$$M - G = G(s - x)^2,$$

where

$$x = 2\frac{s_1 s_2 - 1}{s_1 + s_2}$$

and

$$G = \frac{(s_1 + s_2)^2}{(s_1 - s_2)^2 + 4}.$$

21. Project Suppose that the depth of the river is an unknown constant d. Investigate the problem of determining the depth as well as the mass and location of a point source.

Suppose now that we do not have the means to measure the vertical force of gravity induced by the nugget, but we can detect the *relative* effect of this force at different sites. That is, given observations (x_1, G_1) and (x_2, G_2), we are unable to measure G_1 and G_2, but we can determine which of the following relations holds:

$$G_1 < G_2, \quad G_1 = G_2, \quad G_1 > G_2.$$

The aim of the next three activities is to discover a way of approximating the location (but not the mass) of the isolated source.

22. Problem Suppose a source (s, M) gives rise to observations (x_1, G_1), (x_2, G_2) with $x_1 < x_2$ and $G_1 < G_2$. Show that $x_1 < s$.

23. Problem Develop a method of taking observations that eventually guarantees that observations (x_1, G_1) and (x_2, G_2) are obtained where $x_1 < s < x_2$, that is, so that the source is "trapped" between two observation sites.

24. Problem Given that the source is trapped between two observation sites, develop a method of successively taking further observations in such a way that the source is located between a sequence of pairs of observation sites that "zero in" on the source location. Convince yourself that with persistence it is possible to approximate the location of the source to any desired accuracy by relative gravity measurements.

2.3.3 Notes and Further Reading

For more on the physics of gravity measurements, see H. Robert Burger, *Exploration Geophysics of the Shallow Subsurface*, Prentice-Hall, Englewood Cliffs, NJ, 1992. The material in this module is an expanded and corrected version of part of the material in C. W. Groetsch, "Geophysically motivated inverse problems for the classroom," *International Journal of Mathematical Education in Science and Technology* **26** (1995), pp. 379–388. R. E. Bell's "Gravity gradiometry," *Scientific American*, June 1998, is a popular article on instruments for measuring gravity differentials.

2.4 Splish Splash

Course Level:

Precalculus (Algebra)

Goals:

Relate velocity, distance, and time.

Mathematical Background:

Elementary algebra, quadratic equations

Scientific Background:

Distance, velocity, law of falling bodies

Technology:

Calculator

2.4.1 Introduction

The *scattering* problem is one of the most important inverse problems in the sciences. Its general form is simple to explain: a signal of some type is transmitted, strikes an object (the *scatterer*), and is bounced off the object, or *scattered*.

The scattered signal, which has been affected by the scatterer, is then collected and characteristics of the scatterer are inferred from information contained in the scattered signal. Familiar applications of this idea include radar, sonar, and ultrasonic medical imaging.

In this brief module we discuss a very simple inverse scattering problem. The problem was posed by Sir Isaac Newton in his textbook *Universal Arithmetick*:

> A Stone falling down into a Well, from the Sound of the Stone striking the Bottom, to determine the Depth of the Well.

Here we interpret the direct problem as that of determining the time at which the echo is heard. The inverse problem is to determine a physical characteristic of the well, its *depth*, from one aspect of the reflected signal, the echo time.

2.4.2 Activities

1. Question What physical principle governs the relationship between the depth of the well and the time it takes the dropped stone to reach the bottom of the well? (Neglect air resistance.)

2. Question What is the fundamental difference between the velocity of the falling stone and the velocity of the returning sound signal?

3. Exercise Suppose the well is 50 meters deep. Find the time required for a dropped stone to reach the bottom of the well. (Use $9.8 \text{ m}/\text{sec}^2$ for the acceleration of gravity.)

4. Exercise How long does it take for the sound signal to reach the top of the well in Exercise 3? (Use 331 m/sec for the speed of sound.)

5. Problem Using g for the acceleration of gravity, c for the speed of sound, and d for the depth of the well, find a formula for the time t elapsed from the time the stone is released until the time the splash is heard.

6. Problem Solve the inverse problem of determining the depth d of the well from the elapsed time t between the dropping of the stone and the hearing of the splash. Does the equation determining d have a unique solution? Is there a unique physical solution?

7. Exercise A stone is dropped into a well and 4.2 seconds later the splash is heard. How deep is the well?

8. Problem Show that the depth of the well is always less than $gt^2/2$, where the symbols have the same meaning as in Problem 5. What is the physical reason for this?

9. Calculation If $|x|$ is small, the quantity $\sqrt{1+x}$ is well approximated by the polynomial $p(x) = 1 + x/2 - x^2/8 + x^3/16$. To gauge the quality of this approximation, plot the functions $p(x)$ and $\sqrt{1+x}$ on the same axes for $0 \le x \le 1$.

10. Problem From Problem 8, we know that $gt^2/2$ is an overestimate for the depth of the well. In this problem you are asked to develop a correction term to get a better estimate of the depth. The depth is given by $d = gt^2/2 - \delta$, where $\delta = gt^2/2 - d$. (a) Show that $\delta > 0$. (b) Use the result of Problem 6 and the approximation from Calculation 9 to show that $\delta \approx g^2t^3/(2c)$ if $2gt/c$ is small.

11. Question Does the approximation in Problem 10 seem reasonable for the well in Exercise 7?

12. Exercise Compare the answer obtained in Exercise 7 with the approximation of the depth obtained by using the method of Problem 10.

2.4.3 Notes and Further Reading

The well problem is posed in Isaac Newton's posthumously published elementary textbook *Universal Arithmetick*. Newton's book was originally written in Latin and was translated into English by a "Mr. Ralphson." This is the Raphson of Newton–Raphson fame. The problem appears as Problem L (50), p. 308, of Theaker Wilder's edition of the book, which was published in London in 1769. It is also treated very briefly in N. Ya. Vilenkin's book *Successive Approximation*, Pergamon, Oxford, 1964. George Pólya, in his book *Mathematical Methods in Science* (second printing, Mathematical Association of America, Washington, 1977) provides a more extensive treatment based on infinite series.

2.5 Snookered

Course Level:

Precalculus (Elementary Coordinate Geometry)

Goals:

Explore some very elementary inverse scattering problems.

Mathematical Background:

Coordinate geometry, lines

Scientific Background:

The reflection principle

Technology:

None

2.5.1 Introduction

In this module we treat some very elementary inverse scattering problems suggested by the game of billiards. In this game one is "snookered" (the verb derives from the older game of snooker) if the opposing player skillfully arranges her shot so that a ball is left in line between the cue ball and the target ball. In this case one is forced to make a "bank" shot, that is, the cue ball is driven into a cushion bordering the table surface, avoiding the blocking ball, in such a way that the target ball is struck on the rebound.

The physical principle governing this situation is the *reflection principle*: The angle that the incident path makes with the normal to the reflecting curve (i.e., in this case the cushion) is equal to the angle that the reflected path makes with the normal (see Figure 2.6).The reflection principle is based on an idealized version of reality in which it is assumed that the speed of the ball is constant throughout its path (the path of the ball is assumed to be the same as that of a reflected light ray).

The direct problem of determining the reflected path from knowledge of the impact point and the angle of incidence is then straightforward. But a couple of inverse problems immediately suggest themselves. Given the positions of the cue ball and the target ball, what impact points on the cushion will result in bank shots that strike the target ball? The cushion is a straight line and hence is determined by two parameters (e.g., slope and intercept). Suppose that

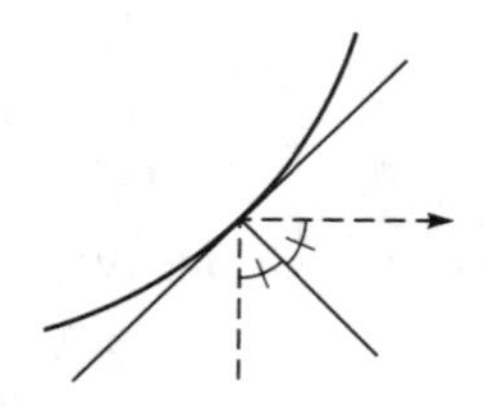

Figure 2.6: The Reflection Principle

the cushion cannot be seen (pool venues have been known to be quite thick with tobacco smoke!) but characteristics of the incident and reflected paths of a bank shot off the cushion can be observed. Is it possible to identify the cushion? These and some related elementary inverse scattering problems are addressed in this module.

2.5.2 Activities

For simplicity let us agree that our "pool table" is the unit square, that is, the set of points

$$\{(x, y) : 0 < x < 1, 0 < y < 1\}$$

in the coordinate plane. We suppose that the cue ball and the target ball are at given distinct positions, say the cue ball is at position (x_1, y_1) and the target ball is at position (x_2, y_2), where $(x_1, y_1) \neq (x_2, y_2)$.

1. Question If $x_1 \neq x_2$ and $y_1 \neq y_2$, how many distinct bank shots will strike the target ball? (By a bank shot we mean a shot that hits a cushion once, then rebounds and strikes the target.)

2. Question If $x_1 = x_2$, how many bank shots result in a hit on the target ball? What about if $y_1 = y_2$?

3. Exercise If the cue ball is at $(.2, .8)$ and the target ball is at $(.7, .6)$, find the impact points on the cushions (remember the cushions are given by $x = 0$ or 1 and $y = 0$ or 1) for which the corresponding bank shots strike the target ball.

4. Problem Find all impact points on the cushion that correspond to bank shots that strike the target ball in Question 1.

5. Problem Suppose the target ball is at $(.8, .5)$ and one wishes to sink it in the corner pocket $(1, 0)$ with a single bank shot. Can this be done if the cue ball is at position $(.2, .3)$?

6. Problem Find all possible positions of the cue ball that will result, after a single bank shot, in sinking the ball in Problem 5 in the corner pocket $(1, 0)$.

7. Question A cue ball is launched from the origin into the first quadrant at an angle θ to the positive x-axis. The ball rebounds off a line and returns to intersect the x-axis at position $r(\theta) > 0$, with the return path making an angle $\psi(\theta)$ with the x-axis (see Figure 2.7). How can the line that caused this scattering behavior be identified?

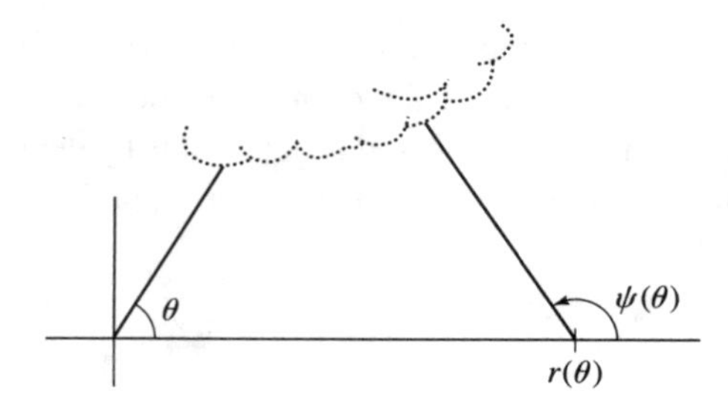

Figure 2.7: Figure for Question 7

8. Question Suppose θ and $\psi(\theta)$ are known in Question 7, but $r(\theta)$ is unknown. Can the scattering line be determined? Can some important characteristic of the scattering line be determined?

9. Exercise Identify the scattering line in Question 7 if $\theta = \pi/6$ radians, $r(\theta) = 10$, and $\psi(\theta) = 3\pi/4$.

10. Problem Suppose a scatterer is an immovable hard circle situated in the first quadrant. Show that two distinct observations of the type described in Question 7 are sufficient to identify the scatterer.

2.5.3 Notes and Further Reading

The reflection principle can be proved (assuming that the speed of the ball is constant) on the basis of *Fermat's Principle*: The time that the ball takes to traverse its path is a minimum. Fermat's Principle is also used to prove a more general result in optics, *Snell's Law*. One can find this derivation in just about any calculus text (see, e.g., G. Thomas, Jr., *Calculus and Analytic Geometry*, Classic Edition, Addison-Wesley, 1983, p. 100). For more on elastic impact, see K. Friedrichs, *From Pythagoras to Einstein*, Random House, 1965.

2.6 Goethe's Gondoliers

Course Level:

Precalculus (Algebra, Geometry)

Goal:

Investigate some elementary inverse problems of location and identification associated with signal timing.

Mathematical Background:

Elementary algebra and geometry, circles, hyperbolas

Scientific Background:

Velocity–distance–time relationships, Fermat's principle of least time

Technology:

Graphing calculator

2.6.1 Introduction

In this module we expand on the theme of the *Splish Splash* module and treat some elementary inverse problems involving the timing of signals. These include a very simple form of the reflection seismology problem and various problems connected with the velocities of sound and light.

We begin with a much simplified model of reflection seismology. Suppose an explosive charge is detonated at some point on the (flat) surface of the earth. A pressure wave then spreads out from the explosive source and propagates through the subsurface. The speed of the wavefront depends on the composition of the subsurface and generally increases with the density of the material. Knowledge of the velocity of propagation can therefore give information about the nature of the material in the subsurface. In our very simple model we assume a homogeneous, horizontally stratified subsurface of constant depth d in which the velocity of propagation of the pressure wave is the constant v. Below the surface layer is hard bedrock off of which the pressure wave is (partially) reflected to eventually arrive again at the surface (see Figure 2.8).

Suppose a receiver, called a geophone, located X units distant from the source detects the reflected signal at a time T after the detonation of the charge. Clearly the relationship between the quantities d, v, X, and T depends on the actual path traced by the ray. The physical principle that determines the path

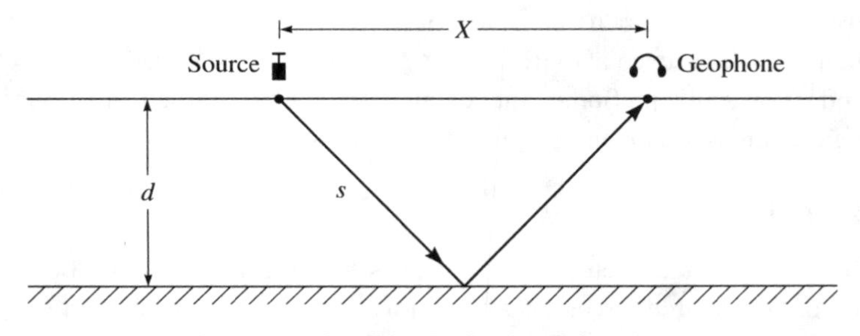

Figure 2.8: Simplified Reflection Seismology

is due to Fermat. Fermat's Principle holds that the travel time along the path must be a minimum. Since we are assuming that the velocity of propagation is constant, this amounts to saying that the total length of the path must be a minimum. A very simple geometrical argument (see Problem 3) shows that Fermat's principle requires that the triangle in the figure must be isosceles. The velocity of propagation is then given by

$$v = \frac{2s}{T},$$

where s is the distance from the source to the point of reflection, and hence by the Pythagorean Theorem,

$$v^2 = \frac{4s^2}{T^2} = \frac{4(d^2 + (\frac{X}{2})^2)}{T^2},$$

or, equivalently,

$$T^2v^2 - 4d^2 = X^2.$$

The *direct* problem of finding the propagation time, T, given the velocity v, the depth d of the stratum, and the known location X of the receiver, then has the unique solution

$$T = \frac{\sqrt{4d^2 + X^2}}{v}.$$

We are more interested in the *inverse* problem of determining the depth d and velocity v from the distance–time observation (X, T). We see from one of the equations derived above that each observation (X, T) gives rise to a *hyperbola*

$$\left\{(d, v) : \frac{v^2}{(\frac{X}{T})^2} - \frac{d^2}{(\frac{X}{2})^2} = 1\right\}$$

consisting of depth-velocity pairs, any of which may be a solution of the inverse problem. The first seven activities below are concerned with various aspects of this reflection problem. Some other elementary inverse problems involving the timing of signals are taken up in the remaining activities.

2.6.2 Activities

1. Question A reflected seismic signal is heard at a geophone 300 meters away from the source 3 seconds after detonation. What is the slowest possible velocity of propagation, irrespective of the depth of the stratum?

2. Question A reflected seismic signal is received 300 meters from the source after 3 seconds. Is it possible for the signal to be received at a geophone 360 meters from the source after 4 seconds?

3. Problem Consider the path from source to geophone that is reflected off the substratum boundary at depth d, as in Figure 2.8. Show that the total length of the path is smallest when the angle that the incident ray makes with the horizontal line at depth d is equal to the angle that the reflected ray makes with the horizontal. Conclude that the triangle in Figure 2.8 is isosceles. (Hint: A simple geometrical argument works: extend the incident ray through the horizontal and consider the point at which the extended ray intercepts the vertical line through the geophone.)

4. Exercise A reflected seismic signal is received 91 meters from the source after 1.1 seconds. The reflected signal is received at a second geophone 200 meters from the source after 2.1 seconds. Find the depth of the stratum (to the nearest meter) and the velocity of propagation (to the nearest meter per second).

5. Calculation Suppose a reflected signal is detected 140 meters from the source after 2.1 seconds. Plot the depth-velocity hyperbola for this observation. If the same signal is heard 400 meters from the source after 5 seconds, plot the depth-velocity hyperbola for this observation. Use the trace feature of your calculator to estimate the depth of the stratum and the velocity of propagation in the stratum.

6. Problem Suppose (X_1, T_1) and (X_2, T_2) are two distance–time observations for the simplified reflection seismology problem satisfying $X_1 < X_2$ and $T_1 < T_2$. Find a condition on these data that ensures the existence of a unique depth d and propagation velocity v for the stratum.

7. Problem For given values of the depth d and velocity v, the travel time T is a function of the horizontal distance X. Comment on the behavior of this travel-time function as X becomes very small or very large. Give physical interpretations of your conclusions.

8. Problem In his *Travels in Italy*, Johann Wolfgang von Goethe made the following diary entry for October 7, 1786, concerning the singing of Venetian gondoliers:

> The singer sits on the shore of an island, on the bank of a canal or in a gondola, and sings to the top of his voice—the people here appreciate volume more than anything else. His aim is to make his voice carry as far as possible over the still mirror of water. Far away another singer hears it. He knows the melody and the words and answers with the next verse. The first singer

answers again, and so on. They keep this up night after night without ever getting tired.

Suppose a listener is positioned on the line segment between two singing gondoliers, A and B, who sing according to Goethe's description (i.e., B is silent while A sings, and B begins to sing immediately upon hearing the last note from A, etc.). One second after hearing the last note from A, the listener hears the first note from B. One and a half seconds after the listener notices that B has stopped singing, he hears the first note of A's reply. How far apart are the gondoliers and where is the listener positioned? (Use 1,100 ft/sec for the speed of sound.)

9. Problem Suppose two people are at different positions, say P and Q, in the plane. Let D be the distance between P and Q. Lightning strikes at another point in the plane. Let t_P be the elapsed time between the instant the observer at P sees the lightning flash and the instant at which he hears the clap of thunder, and let t_Q stand for the corresponding time for the observer at Q. For definiteness, suppose $t_P \leq t_Q$.

(a) Did the lightning strike nearer to P or to Q?

(b) Let r_P be the distance from P to the lightning strike and similarly for r_Q. Is $r_Q + r_P < D$ possible?

(c) Describe the set of possible locations of the lightning strike if $r_P = r_Q$.

(d) Is $r_Q > D + r_P$ possible?

(e) Is $D < r_Q + r_P < r_P + D$ possible?

10. Problem Two people, sitting in their houses, which are 10,000 feet apart, and talking on the telephone, hear the same clap of thunder. One person hears the thunder 4 seconds after the other person. Find the curve consisting of the possible positions of the lightning strike. (Hint: Choose a coordinate system in which the houses are on the horizontal axis and are equidistant from the origin. Use 1,100 ft/sec for the speed of sound.)

11. Problem Take the speed of light to be essentially infinite in comparison to the speed of sound. If in the previous problem the observer nearer to the lightning strike hears the thunderclap 3 seconds after seeing the flash, estimate the possible locations of the strike.

2.6.3 Notes and Further Reading

Timing was at the heart of the primary technological challenge of the eighteenth century: the determination of longitude at sea. Because the earth rotates once in 24 hours, each hour difference in local time between two points on earth corresponds to 360/24 = 15 degrees of longitude. If an accurate shipboard timekeeper could be set to, and maintained at, the time of a fixed reference point, say Greenwich, then the difference between local time at sea and Greenwich time could be used to calculate the longitude of the ship in degrees from Greenwich. The construction of an accurate chronometer that could withstand the heaving of a ship's deck and the rigors of ocean weather was finally achieved by the genius and dogged determination of John Harrison in the mid-eighteenth century.

Prior to Harrison's solution of the longitude problem, William Whiston, Newton's successor as Lucasian Professor at Cambridge and translator of the works of Josephus, and Humphrey Ditton, mathematics master at London's Christ Hospital, suggested a rather desperate method for determining longitude based on the idea of Problem 10. Whiston and Ditton proposed that signal ships be anchored at known positions on a latitude circle. These ships would fire exploding shells high into the sky. A ship could then measure the time difference between the sighting of the flash and the sound of the blast, and thereby determine a circle centered on the signal station on which the moving ship would lie. A second such measurement relative to the next station would place the moving ship on the intersection of two circles. These two intersection points would then determine a unique longitude line.

Accurate determination of position continues to be an important consideration today in both military and civilian applications. The current system that accomplishes this, GPS, the *global positioning system*, works in three dimensions. In GPS, the place of Whiston and Ditton's anchored signal ships is taken by artificial satellites in earth orbit, and the circles and hyperbolas of the activities become spheres and hyperboloids. A nontechnical description of GPS can be found in Strang's article "The mathematics of GPS" in the June 1997 issue of *SIAM News*. *Linear Algebra, Geometry and GPS*, by G. Strang and K. Borre, Wesley-Cambridge Press, 1997, has more on the mathematics of GPS. The stories of John Harrison, and of Whiston and Ditton, are told by Dava Sobel in her book *Longitude*, McGraw-Hill, 1995 (see also D. S. Landes, *Revolution in Time*, Harvard University Press, 1983).

3

Inverse Problems in Calculus

Calculus is a particularly rich source of inverse problems. This should come as no surprise as the inverse relationship between differentiation and integration is at the heart of calculus. Furthermore, as is pointed out repeatedly in the modules, approximate differentiation displays the instability that is the hallmark of inverse problems involving continuous processes. In this chapter, a number of fundamental concepts and techniques from calculus, including the Fundamental Theorem of Calculus, l'Hôpital's rule, Taylor polynomials, the chain rule, improper integrals, and approximate differentiation and integration, are put to use to analyze elementary inverse problems. The continuous processes of calculus also provide the opportunity to develop simple approximation methods based on discretizations of the derivative or integral. A number of MATLAB programs based on such approximate methods, which can be used for numerical explorations of inverse problems, are provided in this chapter.

The applications in this chapter are drawn primarily from mechanics. They include the determination of mass distributions from centroid measurements, the relationship between drain-times and vessel shape, deformations and mass distributions, orbital mechanics, trajectories, and variable interest rates.

3.1 Strange Salami

Course Level:

Calculus (Second Term)

Goals:

Use integrals to model a basic problem in statics. Apply the fundamental theorem of calculus. Investigate notions of existence and uniqueness. Apply technology to analyze an unusual calculus problem.

Mathematical Background:

Integration, differentiation, exponential and logarithmic functions, trapezoidal rule, l'Hôpital's Rule, Fundamental Theorem of Calculus

Scientific Background:

Mass, density distributions, moments, centroids

Technology:

Graphing calculator, MATLAB or other high-level language

3.1.1 Introduction

Consider a bar of length 1 having a nonhomogeneous density distribution. We imagine the central axis of the bar to be the x-axis; the left end of the bar is at the origin and the bar extends to the right. Suppose that the lineal mass density of the bar is a given continuous function f (actually, we can allow for a finite number of jump discontinuities). If $M(x)$ denotes the total mass of the segment $[0, x]$ of the bar, then for a given small increment Δx, the mass of the segment $[x, x + \Delta x]$ is approximately $f(x)\Delta x$, that is,

$$M(x + \Delta x) - M(x) \approx f(x)\Delta x.$$

The precise meaning of the mass density is

$$f(x) = \lim_{\Delta x \to 0} \frac{M(x + \Delta x) - M(x)}{\Delta x} = M'(x),$$

that is, the cumulative mass distribution $M(x)$ is an *antiderivative* of the mass density $f(x)$. The mass of the bar is therefore

$$M(1) = \int_0^1 f(x)\, dx.$$

When we refer to a *density* in this module we will mean a nonnegative function f that is continuous on $(0, 1]$ and positive on $(0, 1)$.

The point at which the bar "balances" is called its *centroid* (or center of mass). The definition of the centroid depends on the concept of *moment*, a notion that is familiar to all who have used a wrench or enjoyed the seesaw as a child. The moment about the origin of a collection of masses $m_1, m_2, \ldots, m_k$ at positions $x_1, x_2, \ldots, x_k$, respectively, on the x-axis is defined to be

$$x_1 m_1 + x_2 m_2 + \cdots + x_k m_k$$

(note that this is a signed quantity). The centroid of this arrangement of masses is defined to be that point $\overline{x}$ such that if an equivalent total mass $M = m_1 + m_2 + \cdots + m_k$ were concentrated at $\overline{x}$, the resulting moment would be the same as that of the original arrangement:

$$M\overline{x} = \sum_{i=1}^{k} x_i m_i.$$

For a continuous bar with mass density f we mimic the approach above by dividing the bar into segments $[x_{i-1}, x_i]$, $i = 1, 2, \ldots, n$, each of length $\Delta x = 1/n$ and approximating the moment of the bar by

$$x_1 f(x_1)\Delta x + x_2 f(x_2)\Delta x + \cdots + x_n f(x_n)\Delta x.$$

The moment of the bar about the origin is then, by definition,

$$\lim_{n\to\infty} \sum_{i=1}^{n} x_i f(x_i)\Delta x = \int_0^1 x f(x)\, dx.$$

Again the centroid $\overline{x}$ is that point at which the concentrated total mass of the bar, $M(1)$, produces an equivalent moment:

$$M(1)\overline{x} = \int_0^1 x f(x)\, dx$$

or

$$\overline{x} = \frac{\int_0^1 x f(x)\, dx}{\int_0^1 f(x)\, dx}.$$

In the same way, we see that for each segment $[0, x]$ of the bar, the centroid of the segment is the position $C(x)$ given by

$$C(x) = \frac{\int_0^x u f(u)\, du}{\int_0^x f(u)\, du}.$$

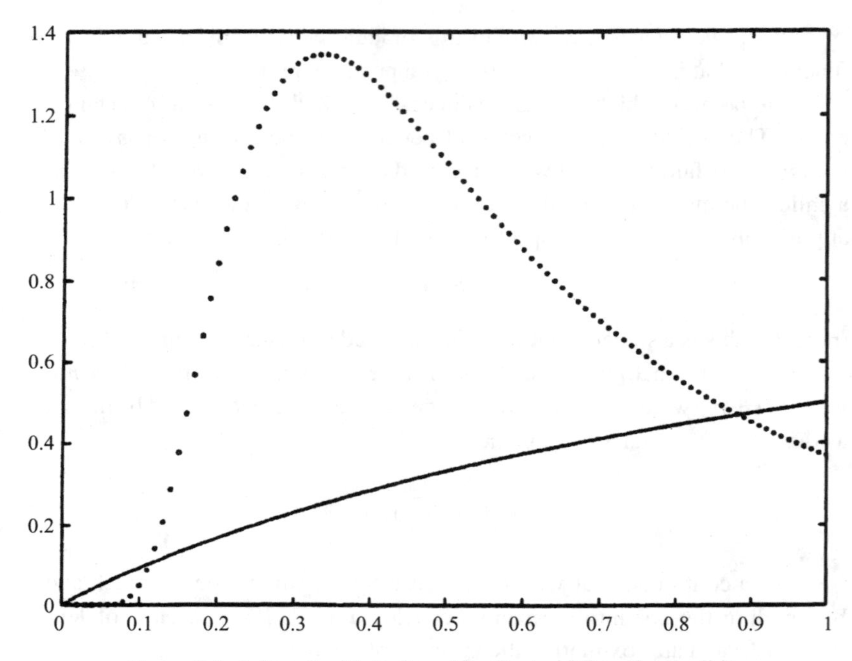

Figure 3.1: A Density (dotted) and Its Centroid Function (solid)

Given the density distribution f, the problem of finding the *centroid function* $C(x)$ is a standard direct problem. The unique centroid function associated with the density distribution is given by the equation above. For example, the density function

$$f(x) = x^{-3}e^{-1/x}$$

plotted in Figure 3.1 gives rise to the centroid function

$$C(x) = \frac{\int_0^x u^{-2}e^{-1/u}\,du}{\int_0^x u^{-3}e^{-1/u}\,du} = \frac{\int_{-\infty}^{-1/x} e^w dw}{-\int_{-\infty}^{-1/x} we^w dw} = \frac{x}{x+1}.$$

3.1.2 Activities

1. Exercise Find the centroid function that corresponds to the density

$$f(x) = 2(x+1)/3.$$

2. Exercise Find the centroid function that corresponds to the density

$$f(x) = 2(x + 1)^{-2}.$$

3. Problem Suppose f is a density and $\{f_n\}$ is a sequence of densities converging (uniformly) to f, that is,

$$|f(u) - f_n(u)| \leq a_n$$

for $u \in [0, 1]$ where $\lim_{n\to\infty} a_n = 0$. Let C be the centroid function generated by f and let C_n be the centroid function generated by f_n. Show that the sequence of centroids $\{C_n(x)\}$ converges to $C(x)$ for each $x \in (0, 1]$.

4. Calculation Plot the centroid function of the density

$$f(x) = 1.1 - e^{-(x-.25)^2}$$

for $0 \leq x \leq 1$. Estimate the centroid of the left-hand quarter of the bar having this density. Do the same for the right-hand quarter of the bar.

5. Problem Show that if C is a centroid function, then $0 < C(x) < x$ for $x > 0$, and give a physical interpretation of this relationship.

6. Problem Find $\lim_{x\to 0^+} C(x)$ and give a physical interpretation.

7. Question On the basis of physical intuition, would you expect $C(x)$ to be an increasing or decreasing function of x?

8. Problem

(a) Show that C is a differentiable function on (0,1) and that

$$C'(x) = \frac{f(x)}{\int_0^x f(u)\,du}(x - C(x)).$$

(b) Use Problem 5 to conclude that $C'(x) > 0$ for $x \in (0, 1)$. Does this square with your answer to Question 7?

9. Problem By $C'(0)$ we mean $\lim_{x\to 0^+} C(x)/x$, assuming this limit exists. (Why is this definition reasonable?)

(a) Show that if $C'(0)$ exists, then $0 \leq C'(0) \leq 1$.
(b) Show that if f is continuous at 0 and $f(0) \neq 0$, then $C'(0) = \frac{1}{2}$.

10. Exercise While $f(x) = x^{-1/3}$ is not a physically realizable density (why?), show that nevertheless the corresponding centroid function $C(x)$ is well-defined.

The remainder of this module is concerned primarily with the inverse problem of determining the density from knowledge of the centroid function. It is helpful to imagine the bar to be a (nonuniformly) stuffed salami, and the challenge to be determining the distribution of the stuffing. Of course, there is a direct way to estimate this density: Simply cut out little slices and weigh them. But, just as the Greeks had their geometrical ground rules concerning the compass and the straight edge, we will also limit the tools that may be used in investigating the inverse problem. We suppose that you have at your disposal only an accurate tape measure and a sharp knife, but no scale.

11. Question How can $C(x)$ be obtained using only the tape measure and the knife?

12. Question Can two different densities give rise to the same centroid function?

Let's explore the uniqueness question a bit more. If f and g are two densities that give rise to the *same* centroid function, then by Problem 8(a) we have

$$\frac{f(x)}{\int_0^x f(u)\,du} = \frac{g(x)}{\int_0^x g(u)\,du},$$

for $x \in (0, 1]$. We would like to see what this implies about f and g.

13. Problem Suppose that f and g are densities that give rise to the same centroid function.

(a) Show that

$$\frac{f'(x)}{f(x)} = \frac{g'(x)}{g(x)}.$$

(b) Conclude that

$$\frac{d}{dx}\ln f(x) = \frac{d}{dx}\ln g(x),$$

and hence

$$f(x) = Dg(x)$$

for some constant $D > 0$.

14. Problem Show that if two bars have differentiable density functions and the same centroid function, as well as the same total mass, then the density distributions of the bars are identical.

15. Problem We now investigate a method of constructing a density, given a centroid function. Suppose C is the centroid function of some density; then by Problems 5, 6, and 8, C satisfies the necessary conditions

$$C(0) = 0, \quad 0 < C(x) < x, \quad C'(x) > 0.$$

Our goal is to use these conditions to construct a density f that gives rise to the centroid function C. In order to impose uniqueness on our solution, we will seek a density with total mass 1 (see Problem 14). The key to the construction is Problem 8(a). Given the centroid function C, let

$$B(x) = \frac{C'(x)}{x - C(x)}$$

and

$$A(x) = \int_0^x f(u)\,du.$$

Then $A'(x) = f(x)$ and the total mass assumption gives $A(1) = 1$. By Problem 8 and the definition of B, we have

$$A'(x) = f(x) = B(x)A(x).$$

(a) Show that

$$\frac{d}{dx} \ln A(x) = B(x).$$

(b) Using the total mass assumption, argue that

$$A(x) = e^{\int_1^x B(u)\,du}.$$

(c) Conclude that the required density is

$$f(x) = A'(x) = B(x)e^{\int_1^x B(u)\,du}.$$

16. Calculation Using the centroid function found in Exercise 1 and the method of Problem 15, construct a density of total mass 1 that gives rise to the centroid function.

In Problem 3 we saw that the direct problem is *stable*: Nearly identical densities give rise to nearly identical centroid functions. It is natural to wonder if the inverse problem exhibits similar stability. That is, if a sequence of centroid functions converges uniformly to a centroid function, do the associated densities converge to the corresponding density? We will see that intuition might not provide a sure guide in this situation. In fact, the next activity shows that the stability of the inverse problem cannot be counted on.

17. Exercise

(a) Let $C(x) = x/2$. Show that $C(x)$ is the centroid function of the density $f(x) = 1$.

(b) For $n = 3, 4, 5, \ldots$, let

$$C_n(x) = \frac{x}{2} + \frac{1}{n}x^{n^2}.$$

Show that $\{C_n\}$ converges uniformly to the centroid function of part (a).

(c) Let f_n be the density with total mass 1 associated with C_n. Show that $f_n(1) \to \infty$ as $n \to \infty$ and hence that $\{f_n\}$ does not converge to f.

18. Problem Find a density that is its own centroid function.

19. Project Consider a two-dimensional density on the unit square $\{(x, y) : 0 \le x \le 1, 0 \le y \le 1\}$. Given a point (x, y) in the square, let

$$C(x, y) = (C_1(x, y), C_2(x, y))$$

be the centroid of the rectangular region $\{(u, v) : 0 \le u \le x, 0 \le v \le y\}$. Carry out as many of the ideas above as you can for this two-dimensional model.

We close this module by suggesting a simple numerical method for approximating a density f with $f(0) > 0$ from given centroid data. We suppose that the values of the centroid function C are known at certain equally spaced points $x_j = jh$, $j = 0, 1, 2, \ldots, n$, where $h = 1/n$. We begin with the relationship

$$C(x) = \frac{\int_0^x uf(u)\,du}{\int_0^x f(u)\,du},$$

which gives

$$C(x_j)\int_0^{x_j} f(u)\,du = \int_0^{x_j} uf(u)\,du.$$

We now approximate each of the integrals by the trapezoidal rule:

$$C(x_j)\frac{h}{2}[f(x_0) + 2f(x_1) + \cdots + 2f(x_{j-1}) + f(x_j)]$$
$$\approx \frac{h}{2}[2x_1 f(x_1) + \cdots + 2x_{j-1}f(x_{j-1}) + x_j f(x_j)].$$

Therefore, for $j = 1, 2, \ldots, n$,

$$f(x_j) \approx \frac{C(x_j)f(0) + 2\sum_{i=1}^{j-1}(C(x_j) - x_i)f(x_i)}{x_j - C(x_j)}.$$

In particular, for "real" densities with $f(0) \neq 0$, we may approximate the normalized density with $f(0) = 1$ corresponding to the given centroid function C by using the method above with the starting value $f(0) = 1$. The MATLAB program 'cent' generates the approximations $\{(x_j, f(x_j))\}_{j=0}^{n}$ for given centroid data.

20. Computation Use the MATLAB program 'cent' with $n = 101$ (i.e., 100 subintervals) to approximate the density f with $f(0) = 1$, which corresponds to the centroid function

$$C(x) = \frac{2x^2 + 3x}{3x + 6}.$$

Plot your approximation and the true density on the same axes.

21. Computation Repeat Computation 20, but use the centroid function

$$C(x) = \frac{(x + 1)\ln(x + 1) - x}{x}.$$

22. Computation Repeat Computations 20 and 21, but perturb the data with uniform random noise of amplitude .0001. Do the same with noise of amplitude .001.

23. Project Develop a computational method to deal with noise in the centroid data.

3.1.3 Notes and Further Reading

A very nice treatment of moments, including their use in motivating the "sign" rules of elementary algebra, can be found in Pólya's *Mathematical Methods in Science*, Mathematical Association of America, Washington, 1977. (See also E. Mach, *The Science of Mechanics*, Open Court, LaSalle, IL, 1942). The inverse centroid problem is discussed in C. W. Groetsch, "Shapes of centroids as inverse problems," *PRIMUS* **3** (1993), pp. 315–322.

3.2 Shape Up!

Course Level:

Calculus

Goals:

Investigate the interplay of direct and inverse problems in a simple drainage model.

Mathematical Background:

Fundamental Theorem of Calculus, improper integrals, change of variables, chain rule

Scientific Background:

Torricelli's Law

Technology:

Graphics–symbolic calculator, MATLAB

3.2.1 Introduction

Suppose a vessel is formed by revolving a curve $x = f(y)$ about the y-axis. The vessel can be filled to various depths with water, and the water is then allowed to flow out under the influence of gravity through an orifice of cross-sectional area a at the base of the vessel. We suppose the vessel itself is hidden from view, being masked, say, by a right circular cylinder of radius and height 1 foot (this of course places an upper bound on the height and diameter of the vessel). However, the depth of water in the vessel can be observed, say, by means of a pilot tube of negligible volume, as shown in Figure 3.2 (an institutional coffee urn of unknown internal geometry is a suitable mental image).

For a given water depth y, the time $T(y)$ that it takes for the vessel to drain completely can be measured with a stopwatch. This "drain-time" function of

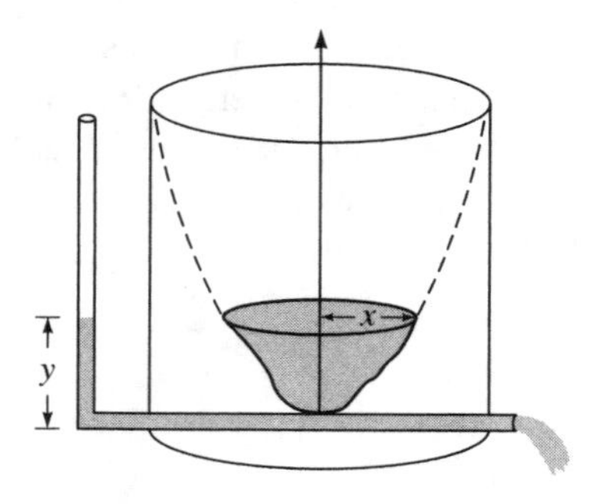

Figure 3.2: A Weird Urn

course depends on the shape, f, of the vessel. In this module we investigate the relationship between T and f.

The rate at which the water flows out of the vessel is given by Torricelli's Law (see the *A Little Squirt* module). According to this law, the velocity of the effluent is $\sqrt{2gy}$, where g is the acceleration of gravity and y is the water depth. In a small increment of time Δt, the change in the volume of water in the vessel, which can be visualized as the volume of a tiny solid cylinder of water of cross-sectional area a that emerges through the orifice in time Δt, therefore satisfies

$$\Delta V \approx -a\sqrt{2gy}\Delta t$$

or

$$\frac{dV}{dt} = -a\sqrt{2gy}.$$

The change in volume can also be computed in terms of the drop in water level Δy. Namely, the volume of water lost when the water level drops by Δy can be visualized as the volume of a tiny circular disk of radius $f(y)$ and height Δy; that is,

$$\Delta V \approx \pi(f(y))^2 \Delta y$$

or

$$\frac{dV}{dy} = \pi(f(y))^2.$$

Therefore,

$$-a\sqrt{2gy} = \frac{dV}{dt} = \frac{dV}{dy}\frac{dy}{dt} = \pi(f(y))^2\frac{dy}{dt},$$

and we find that

$$1 = \frac{-\pi(f(y))^2}{a\sqrt{2gy}}\frac{dy}{dt}.$$

The time $T = T(y)$ required for the vessel to drain if the initial water level is y is then

$$T = \int_0^T 1\,dt = -\int_0^T \frac{\pi(f(y))^2}{a\sqrt{2gy}}\frac{dy}{dt}\,dt$$

$$= -\frac{\pi}{a\sqrt{2g}} \int_y^0 \frac{(f(u))^2}{\sqrt{u}}\, du$$

$$= \frac{\pi}{a\sqrt{2g}} \int_0^y \frac{(f(u))^2}{\sqrt{u}}\, du.$$

This gives the basic relationship between the drain-time function T and the shape function f (we assume that f is continuous on $(0, 1]$ and that $f(u)^2/\sqrt{u}$ is integrable on $[0, 1]$). We will refer to the problem of computing the drain-time function T, given the shape function f, as the *direct* problem. The *inverse* problem consists of determining the shape f given the drain-time function T. This inverse problem may be uniquely solved by using the Fundamental Theorem of Calculus:

$$T'(y) = \frac{\pi}{a\sqrt{2g}} \frac{(f(y))^2}{\sqrt{y}}, \quad \text{for} \quad y > 0,$$

and hence

$$f(y) = \sqrt{\frac{a\sqrt{2g}}{\pi}}\, y^{1/4} \sqrt{T'(y)}.$$

For the remainder of this module we make a specific choice for the cross-sectional area of the orifice, namely

$$a = \frac{\pi}{100\sqrt{2g}}\ \text{ft}^2$$

or just over one-half of a square inch. The relationships between the drain-time function T and the shape f then have the neater forms

$$T(y) = 100 \int_0^y \frac{(f(u))^2}{\sqrt{u}}\, du$$

and

$$f(y) = \frac{1}{10} y^{1/4} \sqrt{T'(y)},$$

where T is measured in seconds and f and y are measured in feet. Recall that we have agreed to the overall constraints $0 \le y \le 1$ and $0 \le f(y) \le 1$ (see Figure 3.2).

The direct problem of determining the drain-time T for any given initial depth is then a straightforward matter of integration, while the inverse problem of determining the shape from the drain times involves differentiation. The relationship between shape and drain time can be put in an even simpler form if we introduce the function

$$F(y) = 100\frac{(f(y))^2}{\sqrt{y}},$$

which we might call the "effective shape," as it determines, and is determined by, the true shape f. The relationship between T and F now takes the simple form

$$T(y) = \int_0^y F(u)\,du, \quad F(y) = T'(y),$$

the classic inverse relationship in calculus.

It is not difficult to bypass the integral and construct a suitable approximation scheme for the drain-time problem. Note that the integral in the relationship

$$T(y) = 100\int_0^y \frac{(f(u))^2}{\sqrt{u}}\,du$$

is *improper*, as the integrand is not defined at $u = 0$. Nevertheless, the function $1/\sqrt{u}$ is integrable, and this fact may be exploited to construct a slightly nonstandard approximate integration routine. Suppose N is a given positive integer, and let $h = 1/N$. We construct an approximation to $T(jh)$ by taking f to be approximately constant on each of the subintervals $[ih, (i+1)h]$, say

$$f(y) \approx f(ih) \quad y \in [ih, (i+1)h], \quad i = 0, 1, \ldots, N-1.$$

This "piecewise constant" approximation to f is illustrated in Figure 3.3.

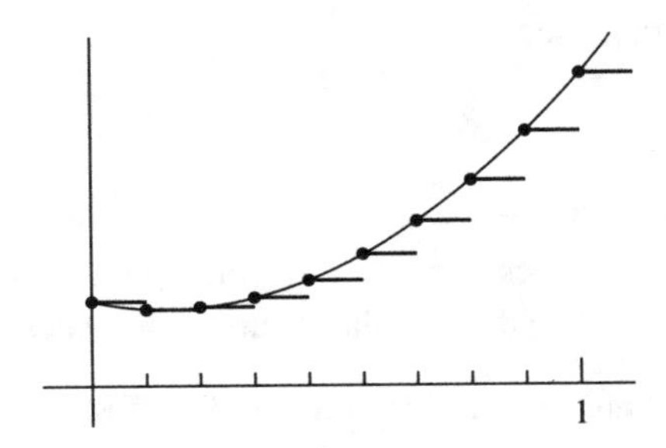

Figure 3.3: Piecewise Constant Approximation

Using this approximation, we have

$$T(jh) - T((j-1)h) \approx 100 \int_{(j-1)h}^{jh} \frac{(f((j-1)h))^2}{\sqrt{u}}\, du$$
$$= 200\sqrt{h}\left(\sqrt{j} - \sqrt{j-1}\right)(f((j-1)h))^2.$$

Since $T(0) = 0$, this gives a very simple recursive strategy for the approximations $T_j \approx T(jh)$:

$$T_0 = 0, \quad T_j = T_{j-1} + 200\sqrt{h}\left(\sqrt{j} - \sqrt{j-1}\right)(f((j-1)h))^2.$$

The program 'drain' computes these approximations, returning a vector of y-values $[0, h, 2h, \ldots, 1]$ and corresponding drain-time approximations $[T_0, T_1, \ldots, T_N]$ for a given positive integer N. The shape function f must be supplied to 'drain' and is defined in a separate routine, 'shape2'. The shape may be perturbed by uniform random error of a given amplitude 'ep'.

A crude approximation scheme for the inverse problem of determining the shape f given the drain-time function T is also easy to come by. Since the effective shape F is simply the derivative of T, we could use the simple approximation

$$F(jh) \approx \frac{T((j+1)h) - T(jh)}{h}, \quad j = 0, 1, \ldots, N-1,$$

where $h = 1/N$. The values of the true shape may then be approximated by

$$f(jh) \approx \frac{(jh)^{1/4}}{10}\sqrt{F(jh)}.$$

The program 'shape' approximates values of the shape given a vector of drain times T. This program also allows the data vector of drain times to be polluted by simulated random noise.

3.2.2 Activities

1. Problem A classic inverse problem in ancient technology was the design of a *clepsydra*, or water clock. One design would have the depth y change at a constant rate with time. What is the shape function of such a clepsydra?

2. Question What are the physical units of the effective shape?

3. Exercise Show that the drain-time function T is increasing.

4. Exercise Find the drain-time function for a right circular cylinder of radius r.

5. Problem Find the drain-time function for a right circular cone of height and radius 1 with vertex at the origin.

6. Problem Show that for any shape of the form $f(y) = y^n$, where $n > 1/4$, the drain-time function is concave up.

7. Problem Construct a shape f for which the drain-time function T has precisely two inflection points.

8. Problem Suppose the drain-time function has the form $T(y) = b(e^{ay} - 1)$ for some positive constants a and b. Find $\lim_{y \to 0^+} f'(y)$.

9. Problem Suppose two shape functions f and $\tilde{f}$ differ by at most ϵ, that is,

$$| f(y) - \tilde{f}(y) | < \epsilon, \quad y \in [0, 1].$$

Let T and $\tilde{T}$ be the drain-time functions corresponding to f and $\tilde{f}$, respectively. Find an upper bound for the difference of drain times $| T(y) - \tilde{T}(y) |$. Show that $\tilde{T}(y) \to T(y)$ for all y as $\epsilon \to 0$.

10. Computation Use the program 'drain' (invocation: $[y, T]$ = drain $(N, `f`, ep);$) to compute the drain-time function for the shape $f(y) = y^2$. Compare the computed approximate solution with the exact solution by plotting both on the same set of axes. Experiment with various values of N. Repeat your experiments using data that has been polluted with random noise of amplitude $ep = .001, .01, .1$. Note the effect of data noise on the computed solution.

11. Computation Use 'drain' to approximate and plot the drain-time function for the shape $f(y) = (1 + \sin(8\pi y))/2$. Add noise of various amplitudes to the data and note the effect of noisy data on the approximations.

12. Problem For $n > 2$, let $T_n(y) = y/2$ for $y \in [0, .5 - 1/n]$,

$$T_n(y) = \frac{p\left(\frac{2ny+2-n}{4}\right)}{2\sqrt{n}} + \frac{y}{2}, \quad y \in [.5 - 1/n, .5 + 1/n]$$

and $T_n(y) = (\sqrt{n}y + 1)/2\sqrt{n}$ for $y \in [.5 + 1/n, 1]$, where $p(x) = -2x^3 + 3x^2$. Show that T_n is differentiable and nondecreasing on $[0, 1]$ and that $T_n(y) \to y/2$ as $n \to \infty$, for all $y \in [0, 1]$. Do the associated shape functions f_n converge to the shape corresponding to $T(y) = y/2$?

13. Computation Use the program 'shape' (invocation: [f]=shape(N,T,ep);), where T is a *vector* of drain times) to approximate the shape whose drain-time function is $T(y) = \sqrt{y^3}$. Use various values of N and plot the approximate shape and true shape in each case. Investigate the effect of data noise on the computed solution.

14. Computation Use the drain times computed in Computation 11 as inputs to 'shape' and compute the corresponding shape function. Plot the computed shape along with the exact shape. Experiment with noisy data and observe the effect on the computed shape.

15. Problem Galileo used a clepsydra to measure small time intervals in his dynamical experiments. The design was a pail of large cross section with a small orifice at its base. Galileo took equal weights (= equal volumes) of effluent to represent equal times (see the Notes). This would be the case only if dV/dt is constant. Show that this is impossible. Suppose, however, that a pan has the form of a right circular cylinder of radius 1 ft and height .4 ft and the area of the orifice is 5×10^{-4} square feet (about the cross section of a pencil). Show that during the first 3 seconds the rate of change of the volume deviates from a constant by less than one-tenth of one percent.

16. Project Suppose that instead of observing the drain-time function $T(y)$, the water depth history $y(t)$ is observed. Develop a theory relating the direct problem of determining $y(t)$ from the shape function $f(y)$ and the inverse problem of determining the shape $f(y)$ from the depth history $y(t)$. Address all of the issues raised in this module and develop software for solving the direct and inverse problem.

3.2.3 Notes and Further Reading

The clepsydra was a marvel of ancient engineering that was more highly developed in the east than in the west. Gibbon reports that the caliph of Baghdad, Harun al-Rashid (764?–809), perhaps in order to appropriately impress Charlemagne (768–814) with the superiority of Islamic science, presented him with a clepsydra along with other valuable gifts such as an elephant and the keys to the Holy Sepulchre! Before the invention of mechanical clocks, a fine clepsydra passed as a precision timekeeper.

Galileo described his method of measuring time for his dynamical experiments in his *Two New Sciences*:

> As to the measure of time, we had a large pail filled with water and fastened from above, which had a slender tube affixed to its bottom, through which

> a narrow thread of water ran; this was received in a little beaker during the entire time that the ball descended along the channel or parts of it. The little amounts of water collected in this way were weighed from time to time on a delicate balance, the differences and ratios of the weights giving us the differences and ratios of the times, and with such precision that, as I have said, these operations repeated time and again never differed by any notable amount.

Other material related to this module can be found in C. W. Groetsch, "Inverse problems and Torricelli's Law," *College Mathematics Journal* **24** (1993), pp. 210–217. Another interesting calculus problem inspired by Torricelli's Law can be found in N. Lord, "A holey unexpected result?," *Mathematical Gazette* **77** (1993), pp. 361–362. For more on Torricelli's work in physical science and mathematics see M. Segre's *In the Wake of Galileo*, Rutgers University Press, New Brunswick, NJ, 1993.

3.3 What Goes Around Comes Around

Course Level:

Calculus (Third Term)

Goal:

Investigate some fundamental inverse problems in celestial mechanics.

Mathematical Background:

Vector calculus, polar coordinates, analytic geometry of conic sections

Scientific Background:

Newton's laws of motion

Technology:

MATLAB or other high-level numerical software

3.3.1 Introduction

In this module some elementary inverse problems in celestial mechanics are addressed. Before we take up the inverse problems we first treat the corresponding direct problems. The topic of this module is the very foundation stone of mathematical science: Newton's theory of motion. Newton laid out this majestic theory in his *Principia Mathematica*, published in 1687. The aim of the *Principia* was nothing less than a rigorous, axiomatic, mathematical treatment of the physical world, à la Euclid. The central theme of the *Principia* is the interplay of force and motion. There are two aspects of this pas de deux: the direct problem of determining the motion, given the force; and the inverse problem of determining the force, given the motion. Newton was quite explicit about the importance of both viewpoints. In the preface to the first edition of the *Principia*, he writes

> I offer this work as the mathematical principles of philosophy, for the whole burden of philosophy seems to consist of this—from the motions investigate the forces of nature, and then from these forces to demonstrate the other phenomena....

Note that Newton places the inverse problem first. Before it is possible to demonstrate "other phenomena" it is necessary to find the laws of nature from the observed motions. And Newton had, thanks to Brahe and Kepler, accurate information on the motions of heavenly bodies. Newton acknowledged that he was able to develop his system of the world because he "stood on the shoulders of giants." Among those giants were Galileo Galilei and Johannes Kepler. From Galileo Newton got the law of falling bodies and (perhaps) the law of inertia: A body that experiences no external force is either at rest or moves with constant velocity along a straight line. Kepler gave the world the three great laws of his new "physical" astronomy:

(1) The orbit of a planet is an ellipse with the sun (the center of attraction) at a focus.

(2) The radius vector from the sun to a planet sweeps out equal areas in equal times.

(3) The square of the period of a planet is proportional to the cube of the length of the major axis of its orbit.

It is important to remember that Kepler's laws are *empirical*; it took Newton's great genius to establish them on the basis of "mathematical principles."

Newton began his mathematical investigation of gravity in 1666. At that time he contemplated the force necessary to hold the moon in its orbit. He took the lunar orbit to be circular of radius r and assumed the speed in the orbit was constant. The speed was then $v = 2\pi r/T$, where T is the period of revolution, and he reasoned that the force on the moon was then v^2/r (see Exercise 1, below). Next Newton supposed that Kepler's third law held for the moon as well as the planets and hence T^2 was proportional to r^3. He then concluded that

$$v \propto \frac{r}{T} \propto \frac{r}{r^{3/2}} = r^{-1/2},$$

and hence $v^2 \propto r^{-1}$. Therefore, the force v^2/r is proportional to r^{-2}. This is the famous inverse-square law. Newton then presumably extrapolated this law from the assumed circular orbit of the moon to the actual elliptical orbits of the planets. Eighteen years later, when Halley visited Newton at Cambridge and asked him what type of orbital curve would be implied by an inverse-square gravitational force, Newton replied immediately, "An Ellipsis." When the stunned Halley asked how he knew this, Newton answered, "I have calculated it...." This meeting eventually led to the publication of the *Principia*, in which Newton made the inverse-square law a cornerstone of his theory and proved the more general result that an inverse-square central force implied a conic section orbit.

The appropriate mathematical framework for the study of motion is vector calculus. The position of a point particle (relative to the usual coordinate system) is given by a vector function $\mathbf{r}$ of time t:

$$\mathbf{r}(t) = x(t)\mathbf{i} + y(t)\mathbf{j} + z(t)\mathbf{k}.$$

Vectors will be symbolized in boldface; the length of a vector will be indicated by the same symbol in ordinary type, e.g.,

$$r(t) = \| \mathbf{r}(t) \| = \sqrt{x(t)^2 + y(t)^2 + z(t)^2}.$$

In this module we often use Newton's "dot" notation to indicate the derivative of a variable with respect to time, e.g.,

$$\dot{\mathbf{r}} = \frac{dx}{dt}\mathbf{i} + \frac{dy}{dt}\mathbf{j} + \frac{dz}{dt}\mathbf{k},$$

while derivatives with respect to nontemporal variables are indicated by the more conventional "d" notation of Leibniz. The derivative of the position with respect to time is the *velocity* and the derivative of the velocity with respect to time is the *acceleration*. We assume that our point particle has unit mass and therefore, by Newton's law of motion, the acceleration $\ddot{\mathbf{r}}$ is identified with the force acting on the particle.

Consider now some direct problems of determining the motion of the particle, given the force that acts on the particle. The simplest of these is the law of inertia: If $\ddot{\mathbf{r}} = \mathbf{0}$, then after two integrations we obtain

$$\mathbf{r}(t) = \mathbf{v}t + \mathbf{b},$$

where $\mathbf{v}$ and $\mathbf{b}$ are constant vectors. If $\mathbf{v} \neq \mathbf{0}$, then the particle travels in a line with constant velocity $\dot{\mathbf{r}}(t) = \mathbf{v}$. On the other hand, if $\mathbf{v} = \mathbf{0}$, then the particle is motionless at position $\mathbf{b}$.

We will model planetary motion by assuming that the planet is the particle and that it is acted upon only by the sun, which we place at the origin. Newton conceived of the sun as an attractor that draws the planet toward itself, causing it to orbit the sun in a characteristic curve. He supposed that the force drawing the planet toward the sun depended on the distance r between the planet and the sun. This idea of a *central* force may be expressed in vector notation in the following way:

$$\ddot{\mathbf{r}} = -f(r)\frac{\mathbf{r}}{r};$$

that is, a central force is directed toward the origin and depends only on the distance from the origin. The simple fact that the force is central has two extraordinary consequences. First, if the force is central, then by simple properties of the cross product we obtain

$$\mathbf{r} \times \ddot{\mathbf{r}} = -\frac{f(r)}{r}\mathbf{r} \times \mathbf{r} = \mathbf{0},$$

and hence,

$$\frac{d}{dt}(\mathbf{r} \times \dot{\mathbf{r}}) = \dot{\mathbf{r}} \times \dot{\mathbf{r}} + \mathbf{r} \times \ddot{\mathbf{r}} = \mathbf{0} + \mathbf{0} = \mathbf{0}.$$

The vector $\mathbf{r} \times \dot{\mathbf{r}}$ is called the *angular momentum* of the particle. This equation says that under the action of a central force angular momentum is conserved, i.e.,

$$\mathbf{r} \times \dot{\mathbf{r}} = \mathbf{c},$$

for some constant vector $\mathbf{c}$. What is the geometrical meaning of this? The vector $\dot{\mathbf{r}}$ is tangent to the path of motion, so the equation above says that the path of motion lies in the plane through the origin, which is perpendicular to $\mathbf{c}$ (an algebraic proof uses the triple product identity: $\mathbf{r}\cdot\mathbf{c} = \mathbf{r}\cdot(\mathbf{r}\times\dot{\mathbf{r}}) = (\mathbf{r}\times\mathbf{r})\cdot\dot{\mathbf{r}} = 0$). So, under the influence of a central force, motion is *planar*. This simplifies both the visualization and the analysis of centrally directed motion.

It is convenient to picture planar motion in a polar coordinate system, that is,

$$\mathbf{r} = r\cos\theta\mathbf{i} + r\sin\theta\mathbf{j}$$

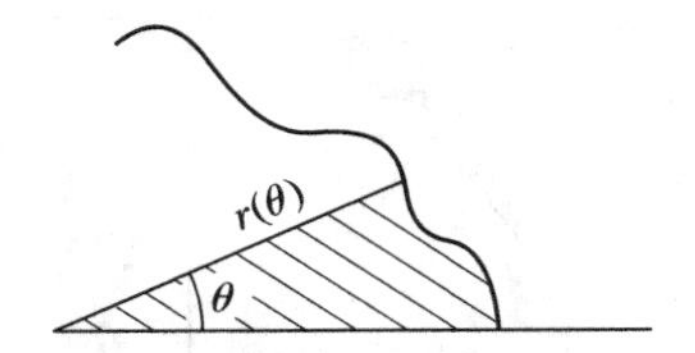

Figure 3.4: Areas in Polar Coordinates

(see Figure 3.4). Consider now the formula for the area A swept out by the radius vector as the angle varies from 0 to θ:

$$A = \frac{1}{2}\int_0^\theta r(\alpha)^2 d\alpha.$$

The *rate* at which area is swept out is then, by the Fundamental Theorem of Calculus,

$$\frac{dA}{dt} = \frac{1}{2}r(\theta)^2\dot{\theta}.$$

However,

$$\dot{\mathbf{r}} = (\dot{r}\cos\theta - r\sin\theta\dot{\theta})\mathbf{i} + (\dot{r}\sin\theta + r\cos\theta\dot{\theta})\mathbf{j},$$

and hence, by conservation of angular momentum,

$$\mathbf{c} = \mathbf{r}\times\dot{\mathbf{r}} = r^2\dot{\theta}\mathbf{k} = \frac{2}{c}\frac{dA}{dt}\mathbf{c},$$

that is, dA/dt is a constant, namely $c/2$.

In other words, under the influence of a central force, the radius vector sweeps out equal areas in equal times. This is Kepler's second law.

Recall that a conic section is characterized as a plane curve, which is the locus of a point the ratio of whose distance from a fixed point O (a *focus*) and from a fixed line L (the *directrix*) is a constant e (the *eccentricity*). An analytical representation of the general conic section in polar coordinates is easy to come by. We put the origin at O and take the polar axis perpendicular to L, as in Figure 3.5. Denote the distance from L to O by k.

The condition $|OP| = e|PQ|$ then becomes

$$r = ek - er\cos\theta,$$

or

$$r = \frac{ek}{1 + e\cos\theta}.$$

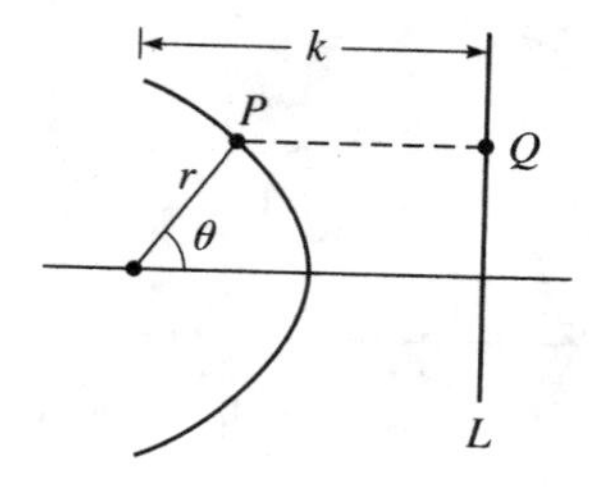

Figure 3.5: A Conic Section

Suppose now that the body orbits under the influence of an inverse-square central force, that is, the position vector satisfies

$$\ddot{\mathbf{r}} = -\frac{a}{r^3}\mathbf{r},$$

where a is a constant. Since the force is central, we have seen that

$$\mathbf{r} \times \dot{\mathbf{r}} = \mathbf{c},$$

where $\mathbf{c}$ is a constant vector. By Problem 3, we find that

$$\frac{d}{dt}\left(\frac{\mathbf{r}}{r}\right) = \frac{\mathbf{c} \times \mathbf{r}}{r^3}.$$

Multiplying by $-a$, we then have

$$-a\frac{d}{dt}\left(\frac{\mathbf{r}}{r}\right) = \mathbf{c} \times \frac{-a\mathbf{r}}{r^3} = -\ddot{\mathbf{r}} \times \mathbf{c}.$$

Integrating this gives

$$a\left(\mathbf{e} + \frac{\mathbf{r}}{r}\right) = \dot{\mathbf{r}} \times \mathbf{c},$$

where $\mathbf{e}$ is a constant of integration. Note that $\mathbf{e} \cdot \mathbf{c} = 0$, and hence $\mathbf{e}$ lies in the orbital plane. If we measure the angular displacement of the orbiting body with respect to the axis in direction $\mathbf{e}$, as in Figure 3.6, then it is easy to see that the orbit is a conic section. Indeed, from the last equation we obtain

$$a(\mathbf{r} \cdot \mathbf{e} + r) = \mathbf{r} \cdot (\dot{\mathbf{r}} \times \mathbf{c}) = (\mathbf{r} \times \dot{\mathbf{r}}) \cdot \mathbf{c} = c^2.$$

But $\mathbf{r} \cdot \mathbf{e} = re\cos\phi$, so

$$r(1 + e\cos\phi) = c^2/a$$

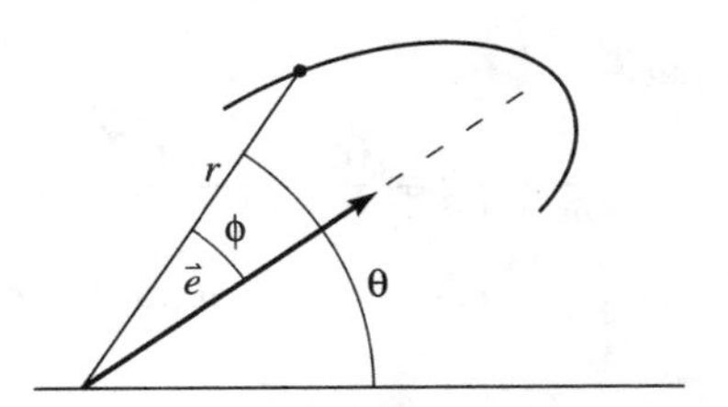

Figure 3.6: Orientation of the Orbit

or,

$$r = \frac{ek}{1 + e\cos\phi},$$

where $k = c^2/(ae)$, which is a conic section of eccentricity e with focus at the center of attraction. Depending on the value of e, the conic may be an ellipse, a parabola, or a hyperbola (see Problem 4). This is Kepler's first law.

It is useful at this point to summarize the main results for a body orbiting a center of attraction:

no force $\Longrightarrow$ uniform straight line motion (or rest)

central force $\Longrightarrow$ planar orbit + equal areas law

inverse square central force $\Longrightarrow$ conic orbit with focus at origin.

What Newton did was provide a firm mathematical justification for these observations of his predecessors (to be sure, his development was strictly Euclidean and looked nothing like the treatment given above). It is natural to ask about the corresponding inverse problems: To what extent may the implications "$\Longrightarrow$" be replaced by "$\Longleftarrow$?" These and other inverse problems were addressed by Newton in the *Principia*. The reader, equipped with the powerful techniques of vector calculus, is invited to investigate a number of such inverse problems in the following activities.

3.3.2 Activities

1. Exercise Show that in a uniform circular orbit

$$\mathbf{r} = r\cos at\mathbf{i} + r\sin at\mathbf{j},$$

the force is central and has magnitude $\|\dot{\mathbf{r}}\|^2/r$.

2. Exercise Verify the vector identity: $(\mathbf{a}\times\mathbf{b})\times\mathbf{c} = (\mathbf{a}\cdot\mathbf{c})\mathbf{b} - (\mathbf{b}\cdot\mathbf{c})\mathbf{a}$.

3. Problem Show that

$$\frac{d}{dt}\left(\frac{\mathbf{r}}{r}\right) = \frac{(\mathbf{r} \times \dot{\mathbf{r}}) \times \mathbf{r}}{r^3}.$$

4. Problem Recall that an inverse-square central force leads to an orbit of the form

$$r = \frac{ek}{1 + e\cos\phi}, \qquad ek = c/a^2.$$

Use more familiar rectangular coordinates to show that if $0 < e < 1$, the orbit is an ellipse, while if $e = 1$, the orbit is a parabola, and if $e > 1$, the orbit is a hyperbola.

5. Question Suppose a particle moves entirely in a plane containing the origin. Is the motivating force necessarily *central*?

6. Problem Give a geometrical proof that for uniform straight-line motion, equal areas are swept out by the radius vector in equal times.

7. Exercise Suppose a body moves in a straight line with constant speed. Show that no force acts on the body.

8. Exercise Use the equal areas law to show that under the influence of a central force,

$$2r\dot{r}\dot{\theta} + r^2\ddot{\theta} = 0.$$

9. Problem Show that a central force implies

$$r\ddot{\mathbf{r}} = (\ddot{r} - r\dot{\theta}^2)\mathbf{r}.$$

10. Problem Show that if a body orbits in a plane containing the origin, and the radius vector sweeps out equal areas in equal times, then the body is motivated by a centrally directed force. (This is Proposition II, Theorem II of Book I of the *Principia*.)

For the remainder of this module, we assume a central force and use the notation $u = r^{-1}$. We assume that the body does not collide with the center of force and hence u is defined for all times.

11. Problem Use the equal areas law to show that $r\dot{\theta}^2 = h^2u^3$, where h is twice the rate at which area is swept out.

12. Problem Show that $\ddot{r} = -h^2u^2\left(\dfrac{d^2u}{d\theta^2}\right)$.

13. Exercise Conclude from the previous three problems that, under the influence of a central force,

$$\ddot{\mathbf{r}} = -g(u)\frac{\mathbf{r}}{r},$$

where

$$g(u) = h^2u^2\left(\frac{d^2u}{d\theta^2} + u\right).$$

This result provides a general tool for the inverse problem for orbits, as the magnitude of the central force is given by $f(r) = g(r^{-1})$. This can be used to find the force functions for a number of simple orbits.

14. Problem Find the central force on a body that orbits on a circular arc that passes through the center of force (*Principia*: Book I, Proposition VII, Corollary I).

15. Problem Show that if a body orbits in a conic with the center of force at a focus, then the force on the body is proportional to the reciprocal of the square of the distance between the body and the center of force (*Principia*, Book I, Propositions XI, XII, XIII).

16. Problem Find the central force on a body whose orbit is a logarithmic spiral, i.e., $r = e^{a\theta}$ (*Principia*, Book I, Proposition IX).

17. Problem Find the central force on a body that orbits in an Archimedean spiral, i.e., $r = a\theta$.

18. Problem Find the central force on a body that orbits in an ellipse whose *center* is the center of force (*Principia*, Book I, Proposition X).

19. Problem Find the central force on a body that orbits on the hyperbola $x^2 - y^2 = a^2$.

20. Problem Find the central force on an object that orbits on the lemniscate $r = a\sqrt{\cos 2\theta}, \quad 0 \le \theta < \pi/4$.

21. Computation Use the program 'ode23' in MATLAB, along with the m-file 'orbit0' provided, to investigate orbits under the constant force law $\ddot{\mathbf{r}} = -\mathbf{r}/r$. Plot the orbits for various initial values. The program is invoked as follows:

$$[t, z] = \text{ode23}('\text{orbit0}', t0, tf, z0),$$

where $t0$, tf are the initial and final values, respectively, of time, $z0$ is the column vector

$$z0 = [x0, y0, \dot{x0}, \dot{y0}]',$$

where $x0$, $y0$ are the initial x and y coordinates, and $\dot{x0}$ and $\dot{y0}$ are the initial velocity components in the x and y directions. The command plot(z(:,1),z(:,2)) will then plot the orbit.

22. Computation Investigate orbits for the inverse square law using ode23 and the m-file 'orbit2'. Use various initial conditions and see if you can produce various conic section orbits.

3.3.3 Notes and Further Reading

The standard modern edition of the *Principia* is *Sir Isaac Newton's Mathematical Principles of Natural Philosophy and His System of the World,* original translation by A. Motte in 1729, revised by F. Cajori, University of California Press, Berkeley, 1946. The most authoritative source on Newton's mathematical work is *The Mathematical Papers of Isaac Newton,* Volumes I–VIII, edited by D. T. Whiteside, Cambridge University Press, Cambridge, 1967–81. Much information of historical importance is found in W. Rouse Ball's *An Essay on Newton's "Principia,"* Macmillan, New York, 1893, and the best overall biography of Newton is R. Westfall, *Never at Rest: A Biography of Issac Newton,* Cambridge University Press, Cambridge, 1980. An elementary treatment of the direct problem, in the inimitable style of Richard Feynman, is contained in D. L. Goodstein and J. R. Goodstein, *Feynman's Lost Lecture: The Motion of Planets Around the Sun,* Norton, New York, 1996. Our treatment of the direct problem follows the terse and elegant presentation of H. Pollard, *Celestial Mechanics,* Mathematical Association of America, Washington, 1976.

Recently a number of valuable works on the history of Newton's dynamics have appeared. Among these are J. B. Brackenridge, *The Key to Newton's Dynamics: The Kepler Problem and the Principia,* University of California Press, Berkeley, 1995, and F. de Gandt, *Force and Geometry in Newton's Principia,* translated by C. Wilson, Princeton University Press, Princeton, 1995. B. Pourciau's "Reading the master: Newton and the birth of celestial mechanics," *American Mathematical Monthly* **104** (1997), pp. 1–19, is an excellent article on the beginnings of dynamics. The May 1994 edition of the *College Mathematics Journal* is a special issue devoted to Newton's work, mainly his work in dynamics. See also Sherman Stein's "Inverse problems for central forces," *Mathematics Magazine* **69** (1996), pp. 83–93.

The inverse problems community takes the position that problems that require finding the cause of an observed effect are inverse problems. Therefore, the problem of finding the force function that gives rise to an observed orbital curve is considered an inverse problem. On the other hand, the history of science community calls this the direct problem, because it was this problem which Newton solved first. The mathematical and the historical communities therefore use directly contradictory terminology, which, not surprisingly, can lead to some confusion. It is interesting to hear what Brackenridge has to say on this (keeping in mind the variance of terminology):

> Building on Newton's description of the nature and universality of the gravitational force, scientists of the eighteenth century shifted their interest almost exclusively from direct [inverse] to inverse [direct] problems. They used the combined gravitational forces of the sun and the other planets to predict and explain perturbations in the conic paths of the planets and comets. That interest continued through the nineteenth and twentieth centuries, and today scientists still concentrate on the inverse [direct] problem—rather than the direct [inverse] one. In fact, the early terminology is often reversed by modern scientists; what was then called the inverse problem is now called the direct problem because it is seen as a direct application of the law of universal gravitation to a particular physical problem. For Newton, however, the challenge of finding the force functions remained the primary and direct problem.

Finally, we should say a few words about Newton's intellectual debt to Robert Hooke. In the introductory chapter we mentioned the bitter dispute between Newton and Hooke over the priority for the inverse-square law. This acrimonious episode led Newton to expunge Hooke's name from the *Principia*. However, a very good argument can be made that Newton got from Hooke the basic idea of a central force in the sun deflecting a planet's motion from its straight-line inertial path and causing it to "fall" toward the sun on an elliptical orbit. Indeed, Hooke gave a fairly explicit description of this process in 1674 (quoted by R. Weinstock in the special issue of the *College Mathematics Journal* mentioned above):

> [A]ll bodies whatsoever that are put into a direct and simple motion, will so continue to move forward in a streight line, till they are by some other effectual powers deflected into a Motion, describing a Circle, Ellipsis, or some other more compounded Curve Line... these attractive powers... so much the more powerful... by how much the nearer the body wrought upon is to their own Centers.

3.4 Hanging Out

Course Level:

Calculus (Third Term)

Goal:

Investigate the interplay of direct and inverse problems for hanging cable models.

Mathematical Background:

Fundamental Theorem of Calculus, iterated integrals, arc length, Taylor polynomials, hyperbolic functions

Scientific Background:

Weight distributions, balancing of forces

Technology:

Graphing calculator with "solve" feature

3.4.1 Introduction

Consider a cable or rope that supports a horizontally distributed load, as for instance in a suspension bridge (see Figure 3.7). We assume that the weight of the cable itself is negligible in comparison to that of the load bed, as is the weight of the vertical "hanger" lines that connect the cable to the horizontal load bed. The weight of the load bed is distributed (nonuniformly) along the interval $-1 \leq x \leq 1$ of the x-axis. Furthermore, we assume that the cable has a unique lowest point, which, in the simple model we shall treat, is assumed to occur at $x = 0$.

Suppose the horizontally distributed load is variable and given in terms of a weight distribution $w(x)$, that is, $\int_{-1}^{x} w(u)\,du$ is the total weight of the segment $[-1, x]$ for each $x \in [-1, 1]$. A given weight distribution w engenders a shape

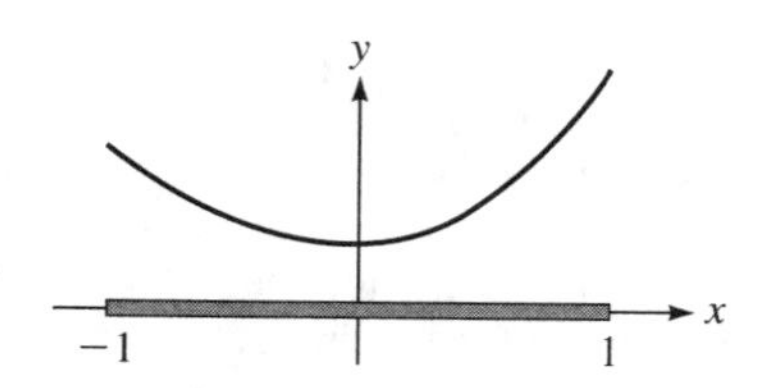

Figure 3.7: A Loaded Cable

$y = y(x)$ for the hanging cable. We consider the problem of finding a shape y caused by a given weight distribution w to be a *direct* problem. The shape of the cable is in principle easily observed, while determining the weight density would generally involve dissecting, and hence destroying, the "road bed." We therefore consider the problem of determining a density distribution w from observation of the shape y to be an *inverse* problem.

The relationship between the weight distribution and the shape is easily obtained by balancing forces. We choose the vertical axis of our coordinate system to pass through the lowest point of the cable (assumed to be weightless), and we imagine the weight of the load to be distributed along the x-axis (connected by invisible vertical lines to the cable) according to a weight distribution $w(x)$. Let H be the horizontal tension at $x = 0$ and T be the tension at a point (x, y), as in Figure 3.8. If θ is the angle between the tangent line at (x, y) and the horizontal, then balancing the component forces we find that

$$T \sin\theta = \int_0^x w(u)\, du$$

and

$$T \cos\theta = H.$$

Since $dy/dx = \tan\theta$ at the point (x, y), we get the basic equation relating the shape and weight density for the loaded cable:

$$\frac{dy}{dx} = \frac{1}{H}\int_0^x w(u)\, du.$$

Note that solving the *direct* problem, that is, determining a shape y from a given weight distribution w, involves solving a *differential* equation, while the solution of the *inverse* problem, that is, determining a weight distribution w given a shape y, entails the solution of an *integral* equation. We refer to this model as the *loaded cable* model; some aspects of direct and inverse problems related to the loaded cable model are investigated in the activities below.

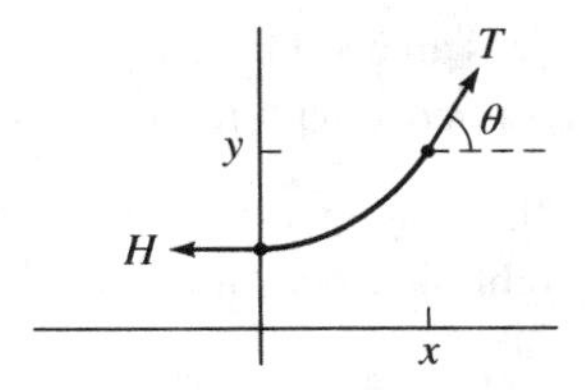

Figure 3.8: Force Balancing

The *free cable* may be investigated using a similar, but more complicated, model. In the free cable model, we assume that the cable hangs under the influence of its own weight (assumed to be, in general, nonuniformly distributed). Here the weight distribution $w(s)$ is taken to be a function of the length s of the arc of the cable from its lowest point. The appropriate model is then, by the same reasoning as above,

$$\frac{dy}{dx} = \frac{1}{H}\int_0^{s(x)} w(u)\,du,$$

where

$$s(x) = \int_0^x \sqrt{1 + y'(t)^2}\,dt$$

is the length of the arc of the cable over the interval from 0 to x. Differentiating with respect to x, we see that the shape y and the weight density w of the cable are related by

$$\frac{d^2y}{dx^2} = \frac{1}{H}w(s(x))\sqrt{1 + y'(x)^2},$$

where H is the horizontal tension at $x = 0$.

3.4.2 Activities

The first group of activities is concerned with the loaded cable model.

1. Question If the weight distribution w and the horizontal tension H are known, is the shape $y(x)$ uniquely determined? What additional condition will allow unique determination of the shape?

2. Problem Show that

$$H = \frac{1}{y(1) - y(0)}\int_0^1 w(u)(1 - u)\,du.$$

3. Problem Show that if w is strictly positive, then y is concave-up.

4. Problem Show that if w is an even function, i.e., $w(-x) = w(x)$, then any corresponding shape y is symmetric with respect to the y-axis.

5. Problem Show that if the shape y is symmetric with respect to the y-axis, then any corresponding weight distribution w is an even function.

6. Exercise Show that $y(x) = x^4$ and $y(x) = 3x^4 + 7$ are both shapes engendered by the weight distribution $w(x) = \frac{3}{2}x^2$.

7. Problem Show that if y is a shape engendered by a weight distribution w then so is $ay + b$ for all positive numbers a and b.

8. Problem We say that a shape y is *normalized* if $y(0) = 0$ and $y(1) = 1$. Show that the unique normalized shape engendered by a given weight distribution w is

$$y(x) = \frac{\int_0^x w(u)(x-u)\,du}{\int_0^1 w(u)(1-u)\,du}.$$

9. Problem Show that if w is a weight distribution that engenders a shape y, then so is aw for all $a > 0$.

10. Problem Show that if w is a weight distribution of total weight 1 (i.e., $\int_{-1}^1 w(u)\,du = 1$), and y is a corresponding shape, then

$$H = \frac{1}{y'(1) - y'(-1)}.$$

11. Problem Show that there is a unique weight distribution of total weight 1 that can account for a given shape y, namely

$$w(x) = \frac{y''(x)}{y'(1) - y'(-1)}.$$

12. Exercise Find the weight distribution with total weight 1 that engenders the shape $y(x) = 1 - \cos(\pi x/2)$.

13. Problem Show that if the weight distribution is constant, then the shape is parabolic.

14. Problem Show that if the shape is parabolic, then the weight distribution is constant.

15. Calculation Find, and plot, the normalized shape engendered by the weight distribution

$$w(u) = \begin{cases} 1 + u & \text{if } -1 \le u < 0 \\ 1 + u^2 & \text{if } 0 \le u < 1 \end{cases}.$$

16. Calculation Find the weight distribution, with total weight 1, which engenders the shape

$$y(x) = \begin{cases} x^2 & \text{if } x < 0 \\ x^4 + x^2 & \text{if } x \ge 0 \end{cases}.$$

17. Problem Let $\{w_n\}$ be a sequence of positive weight distributions and suppose w is a positive weight distribution with

$$|w_n(x) - w(x)| \leq a_n \quad \text{for all } x \text{ with } -1 \leq x \leq 1,$$

where $a_n \to 0$ as $n \to \infty$. Let y_n be the normalized shape engendered by w_n and let y be the normalized shape engendered by w. Show that $y_n(x) \to y(x)$ as $n \to \infty$ for each x.

18. Problem Show that the sequence of normalized shapes

$$y_n(x) = \frac{\cosh(nx) - 1}{\cosh(n) - 1}$$

is engendered by the sequence of weight distributions

$$w_n(x) = \frac{n^2}{\cosh(n) - 1} \cosh(nx).$$

Also show that

$$y_n(x) \to 0 \quad \text{as} \quad n \to \infty \quad \text{for} \quad x \in (-1, 1),$$

while

$$w_n(x) \to \infty \quad \text{as} \quad n \to \infty \quad \text{for} \quad x \neq 0.$$

19. Problem Suppose $y(0) = 0$, where y is a shape in the simplified loaded cable model. Show that the x-intercept of the tangent line to the curve y at the point $x > 0$ is the centroid $C(x)$ of the segment $[0, x]$ of the load bed.

The remaining activities in this module are concerned with the free cable model. We will consider only the simplest case, where the weight distribution is constant, say $w(s) = w$. The governing equation for this constant weight distribution model is therefore

$$\frac{d^2y}{dx^2} = \frac{w}{H}\sqrt{1 + (y'(x))^2}.$$

20. Problem Show that no parabolic shape satisfies the free cable model for a constant weight distribution.

21. Problem Show that the shape $y(x) = a\cosh(x/a)$ satisfies the free cable model for a constant weight distribution w, where $a = H/w$.

22. Problem Show that if $y(x) = a\cosh(x/a)$ satisfies the general free cable model, then the weight distribution w is constant.

We continue to investigate the shape $y(x) = a\cosh(x/a)$ for $-1 \leq x \leq 1$. We refer to the number h defined by

$$h = y(1) = a\cosh(1/a)$$

as the "height" of the shape.

23. Calculation Show that, no matter what the value of $a > 0$, the height is at least 1.5. More precisely, show that $h \geq \sinh x_0$ where $x_0 = \coth x_0$.

24. Exercise Sketch the shape $y(x) = a\cosh(x/a)$ for $-1 \leq x \leq 1$ and show that $h = y(1) > a$.

We define the "sag," s, of the shape $y = a\cosh(x/a)$ to be $s = h - a$.

25. Problem Given a sag s with $0 < s$, show that the equation $sx + 1 = \cosh(x)$ has a unique positive solution, which we shall call $r(s)$.

26. Calculation Estimate $r(s)$ for $s = .1, .5, 1, 2, 4, 10$.

27. Problem Show that $r(s) \to 0$ as $s \to 0^+$, and $r(s) \to \infty$ as $s \to \infty$. Also show that $r(s)$ is an increasing function of s.

28. Problem Suppose the constant weight distribution of a freely hanging cable is .2 lb/ft, for $-1 \leq x \leq 1$ and sag $s = .3$ ft. What is the horizontal tension H?

29. Problem Find the second-degree Maclaurin polynomial $p_2(x)$ for the shape $y(x) = a\cosh(x/a)$.

30. Problem Show that

$$\max_{-1\leq x\leq 1} \left|a\cosh\left(\frac{x}{a}\right) - p_2(x)\right| \to 0, \quad \text{as } a \to \infty.$$

31. Calculation Compare the graphs of $y = a\cosh(x/a)$ and $p_2(x)$ on $[-1, 1]$ for $a = 1, 2, 3$.

3.4.3 Notes and Further Reading

Galileo believed that the shape of a freely hanging cable with uniform weight distribution was a parabola. He evidently came to this conclusion by making an analogy between the cable and a projectile (see the module *A Cheap Shot*). The vertically acting weight of the cable was compared with the vertical force of gravity acting on the projectile and the horizontal tension was (incorrectly) related to the horizontal motion of the projectile. The analogy is drawn by

Galileo's character Sagredo (see *Two New Sciences*, translated by Stillman Drake, Wall and Thompson, Toronto, 1969, p. 256):

> In drawing the rope, there is the force of that which pulls it horizontally, and also that of the weight of the rope itself, which naturally inclines it downward. So these two kinds of events are very similar.

Galileo also suggested that this idea could be used to trace parabolas. His character Salviati describes the method (ibidem, p. 143):

> The other way to draw... the line we seek is to fix two nails in a wall in a horizontal line.... From these two nails hang a fine chain.... This chain curves in a parabolic shape, so that if we mark the points on the wall along the path of the chain, we shall have drawn a full parabola.

The shape assumed by such a chain came to be known as a *catenary* (from the Latin "catenarius" for chain). The term was apparently first used by teenaged Christian Huygens in a letter to Marin Mersenne in 1646 in which Huygens showed, contrary to Galileo's claim, that the catenary is *not* a parabola (See Problem 20, above. Note that for high tensions (Problem 30), Galileo was not far wrong.). In the May 1690 issue of *Acta Eruditorum*, Jacob Bernoulli issued a challenge to the mathematical community to find the true shape of the catenary. The problem was solved by Huygens, Leibniz, and Jacob's brother Johann, among others.

Problem 19 is a special case of a theorem of the Jesuit priest Ignace-Gaston Pardies (1636–1673). Pardies' work was the basis of some early solutions of the catenary problem.

For a treatment of catenary structures from an engineering perspective, see H. M. Irvine, *Cable Structures*, MIT Press, Cambridge, 1981, and P. Broughton and P. Ndumbaro, *The Analysis of Cable and Catenary Structures*, American Society of Civil Engineers, New York, 1994.

3.5 Two Will Get You Three

Course Level:

Calculus (Third Term)

Goals:

Investigate two inverse problems for trajectories.

Mathematical Background:

Chain rule, differentiation of vector functions

Scientific Background:

Newton's laws of motion

Technology:

None

3.5.1 Introduction

In a previous module we discussed Galileo's use of his law of falling bodies to deduce that a projectile in a resistanceless medium follows a parabolic trajectory. The law of falling bodies is in turn equivalent, via Newton's law of motion, to constant gravity. Indeed, if a unit mass is released from rest at a height h, then

$$y = -\frac{g}{2}t^2 + h \quad \text{if and only if} \quad \ddot{y} = -g.$$

So, Galileo derived his parabolic projectile path from the law of falling bodies (equivalently, constant gravity) and lack of resistance. Or did it go the other way around? Could it be that the parabolic trajectory came first and the law of falling bodies was a consequence? The great physicist Ernst Mach speculated on this possibility in his book *The Science of Mechanics*:

> In the riper and more fruitful times of his residence in Padua, Galileo dropped the question of "why" and inquired the "how" of the many motions which can be observed. The consideration of the line of projection and its conception as a combination of a uniform horizontal motion and an accelerated motion of falling enabled him to recognize this line as a parabola, and *consequently* [my italics] the space fallen through as proportional to the square of the time of falling.... Now, whether Galileo attained to the knowledge of the uniformly accelerated motion by consideration of the parabola of projection or in another way, we cannot doubt that he tested the law experimentally as well.

In this mini-module we revisit the problem of the trajectory of a point projectile with respect to a flat earth. For variety, we will consider a projectile of unit mass that is fired horizontally with a given velocity v from a height h as in Figure 3.9: As before, it is a simple matter to deduce that the path of the projectile is parabolic, assuming no resistance and a constant gravitational force that acts vertically (see Exercise 1). That is,

$$\text{no resistance} + \text{constant gravity} \Rightarrow \text{parabolic trajectory}.$$

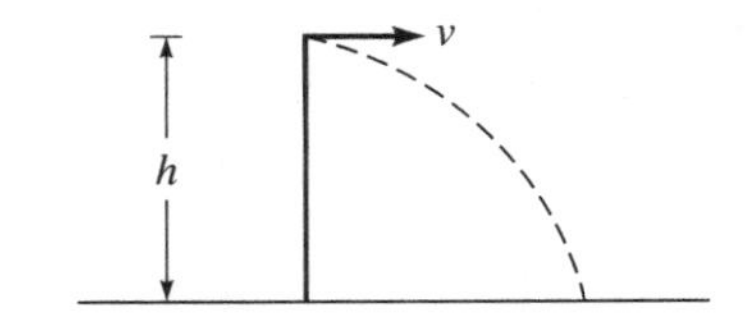

Figure 3.9: A Horizontal Shot from Height

Two inverse problems are immediately suggested: If the trajectory is parabolic and there is no resistance, is the gravitational force necessarily constant? If the trajectory is parabolic and the gravitational force is constant, can there be any resistance? In this unit the reader is led to conclude that any two of the conditions

$$\{\text{no resistance, constant gravity, parabolic trajectory}\}$$

implies the third.

3.5.2 Activities

1. Exercise Suppose a point projectile of unit mass is launched horizontally from a height h with initial velocity v as in the figure above. If the projectile experiences no resistance and gravity acts vertically with constant acceleration g, show that the trajectory is a parabola.

2. Question What are the vertex and focus of the trajectory in Exercise 1?

3. Exercise Show that if the trajectory of a projectile launched as above is parabolic, then its path has the form $y = ax^2 + h$.

4. Exercise Show that if a particle follows the parabolic trajectory of Exercise 3, then the vertical acceleration of the particle at position (x, y) is $2a(\dot{x}^2 + x\ddot{x})$. If the medium offers no resistance, show that the vertical acceleration is constant.

5. Problem Suppose that a particle launched as above follows a parabolic trajectory. Assuming that it experiences a resistance to its motion that is a function of its position and velocity, say, $f(\mathbf{r}, \dot{\mathbf{r}})$, where $\mathbf{r}$ is the position vector of the particle, show that if the vertically acting gravitational force is constant, then there is in fact no resistance (i.e., $f(\mathbf{r}, \dot{\mathbf{r}}) = 0$).

6. Problem Suppose a particle of unit mass follows the hyperbolic trajectory $xy = c$, a constant, and is acted upon by a vertical gravitational force $g(y)$. Find the form of the force $g(y)$.

7. Problem Suppose a particle of unit mass is acted upon by a vertical gravitational force $g(y)$ and follows the arc of a circular trajectory $x^2 + y^2 = a^2$. Find the form of the force $g(y)$.

3.5.3 Notes and Further Reading

Newton, in Book III of the *Principia*, used a model of projectiles fired from a high mountain on a circular earth, similar to that treated in this unit, to explain the motion of orbiting satellites:

> That by means of centripetal forces the planets may be retained in certain orbits, we may easily understand, if we consider the motion of projectiles...

See Cajori's edition of Motte's English translation of the *Principia*, University of California Press, Berkeley, 1946, p. 551. For additional characterizations of parabolic trajectories, see M. S. Klamkin, Problem 88-5, SIAM Review **30** (1988), p. 125.

Problem 7 appears as Proposition VIII of Book I of the *Principia*.

3.6 Uncommonly Interesting

Course Level:

Calculus

Goals:

Investigate a simple inverse problem in finance.

Mathematical Background:

Derivatives, integrals, difference quotients, trapezoidal rule, Taylor's Theorem

Scientific Background:

None

Technology:

Graphing calculator, MATLAB or other high-level numeric language

3.6.1 Introduction

In common interest models (i.e., continuously compounded interest models), the percentage rate of change of an investment is a given constant interest rate r. That is, the rate of change of the value, relative to the current value, is a constant r. To put it mathematically,

$$\frac{du}{dt}/u = r,$$

where u is the *value history*, that is, $u(t)$ is the value of the investment at time t. This, of course, is equivalent to

$$\frac{d}{dt}\ln u = r,$$

which leads to the exponential growth model for the value history:

$$u(t) = u(0)e^{rt}.$$

In the variable interest rate model, the interest rate is time-dependent: $r = r(t)$. The basic relationship

$$\frac{d}{dt}\ln u = r(t)$$

continues to hold, leading to

$$u(t) = u(0)e^{\int_0^t r(\tau)d\tau}.$$

Given the variable interest rate, and the initial value $u(0)$, we regard the problem of determining the value history u as a *direct* problem. On the other hand, finding the interest rate function from the value history is an *inverse* problem. Note that the solution of the direct problem involves integration, a stable process, while the solution of the inverse problem employs differentiation, an unstable process.

Our primary concern is with the inverse problem. If the value history u is known analytically, then finding r is a simple matter of differentiation. We can also *approximate* r given values of u at discrete points in time, $t_k = kh$, $k = 0, 1, 2, \ldots$ where $h > 0$ is a given *time step*. The most straightforward way of doing this is to replace the derivative by a difference quotient. If values $u_k \approx u(t_k)$ are given, then we can approximate $r(t_k)$ by the number

$$r_k = \frac{\ln u_{k+1} - \ln u_k}{h},$$

obtained by replacing the derivative above by the forward difference quotient. The program *rate* computes the approximation to r by this "derivative" method. It also allows the possibility of corrupting the value history with random noise to investigate the stability of the process.

Another approach to approximating the solution of the inverse problem can be developed from the original model. Namely, if we integrate each side of the equation

$$\frac{du}{dt} = ru$$

from t_k to t_{k+1}, we obtain

$$u(t_{k+1}) - u(t_k) = \int_{t_k}^{t_{k+1}} ru\,dt.$$

If the right-hand side is approximated by the trapezoidal rule, and the resulting approximations for $u(t_k)$ and $r(t_k)$ are denoted u_k and r_k, respectively, then we are led to the following equations relating the approximations:

$$u_{k+1} - u_k = \frac{h}{2}(r_k u_k + r_{k+1}u_{k+1}),$$

or

$$r_{k+1} = \frac{2(u_{k+1} - u_k)/h - r_k u_k}{u_{k+1}}.$$

In all the Activities below, the interest rates and value histories are assumed to be strictly positive functions.

3.6.2 Activities

1. Exercise Find the variable interest rate corresponding to the value history $u(t) = \sqrt{t + 5e^{.02t}}$.

2. Problem Show that two value histories have the same interest rate if and only if their ratio is a positive constant.

3. Problem Show that if r is increasing, then u is concave-up.

4. Calculation Suppose a value history u is generated by the interest rate $r(t) = e^{-(t-1)^2}$. Estimate the inflection point of u.

5. Problem Suppose $\{r_n\}$ is a sequence of interest rates converging to a strictly positive interest rate r in the sense that $|r(t) - r_n(t)| \leq a_n$ for all $t \in [0, T]$, where $a_n \to 0$ as $n \to \infty$. Let u_n be the value history generated by the interest rate r_n and u be the value history generated by r. Suppose that $u_n(0) = u(0)$ for all n. Show that $u_n(t) \to u(t)$ for all $t \in [0, T]$.

6. Problem Suppose $u_n(t) = e^{e^{nt}/1+e^{nt}}$. Show that $|u_n(t)| \leq e$ for all t and n, but $r_n(1/n) \to \infty$.

7. Problem Show that if $\ln u$ is concave-down, then the derivative approximation of the interest rate is a *lower estimate* of the true rate (i.e., $r_k < r(kh)$ for $k \geq 1$).

8. Problem Suppose approximations u_k to $u(t_k)$ are given satisfying

$$|\ln u_k - \ln u(t_k)| \leq \delta,$$

where δ is a given positive error bound. Show that if a time step of the form $h = c\sqrt{\delta}$, where c is a positive constant, is used, then the "derivative" approximation r_k to the interest rate satisfies $|r_k - r(t_k)| \leq M\sqrt{\delta}$, for some constant M.

9. Computation Use the program 'rate', with a term of 30 years, and $n = 100$ periods, to estimate the rate that corresponds to the value history in Exercise 1. Estimate the rate using the "derivative" method (but save the results that 'rate' returns using the "integral" method for use in the next activity). Plot the estimated rate and the true rate function and compare. Begin with no noise ($ep = 0$) and repeat the experiment with noise levels of $ep = .001, .01, .05$.

10. Computation Use the results from the previous activity to compare the estimate of rate (with no noise and with the same noise levels as in the previous computation) obtained by the "integral" method with the true rate.

11. Question With no noise in the data, which method ("derivative" or "integral") appears to be more accurate? Which method appears to be more unstable? Can you account for any difference?

3.6.3 Notes and Further Reading

For more on the instability of derivative approximations addressed by Activities 9–11 above, see, for example, C. W. Groetsch, "Differentiation of approximately specified functions," *American Mathematical Monthly* **98** (1991), pp. 847–850 and "Lanczos' generalized derivative," *American Mathematical Monthly* **105** (1998), pp. 320–326.

4

Inverse Problems in Differential Equations

Most "real" inverse problems in science and technology are connected with differential equations. The more interesting problems of this type are usually associated with partial differential equations, but a number of challenging elementary inverse problems arise in ordinary differential equations. This chapter deals with a collection of inverse problems for ordinary differential equations relating to mixing problems, descent problems, hydraulics, heat exchange, and dynamical systems. Also discussed are a few inverse problems for a very basic partial differential equation, the heat equation.

The physical background required in the models includes Newton's laws of motion, Newton's law of cooling, Fourier's law of heat flow, Hooke's Law, and Torricelli's Law. Among the mathematical tools required are solution methods for linear ordinary differential equations (first-order, and second-order with constant coefficients), Laplace transforms, and finite difference methods. In addition, at various points in the modules, basic versions of some fundamental mathematical ideas are introduced without much fanfare, including fixed point analysis, Gaussian integration, and finite elements.

4.1 Stirred, Not Shaken

Course Level:

Differential Equations

Goals:

Investigate some elementary inverse problems in "mixing."

Mathematical Background:

Linear differential equations

Scientific Background:

Relationships between quantity, volume, concentration, and flow rate

Technology:

MATLAB or other high-level numerical language, calculator

4.1.1 Introduction

Mixing problems for well-stirred solutions are standard fare in an elementary differential equations course. Typically the *direct* problem is treated: given an initial concentration of a solute and certain inflow and outflow rates, calculate the concentration at future times. In the simplest model of this type, a vessel has a known volume V, liquid with a given concentration a of solute enters at a given rate r, and the well-stirred solution drains from the vessel at the same rate r.

The construction of the model simply relies on rate balancing. If $q(t)$ represents the quantity of solute in the vessel at time t, then the rate at which q changes with time is the difference between the rate at which the solute enters the vessel (the inflow rate) and the rate at which it leaves (the outflow rate). Therefore,

$$\frac{dq}{dt} = ar - \frac{q}{V}r,$$

or, in terms of the concentration $c(t) = q(t)/V$ of solute in the vessel, the differential equation is

$$\frac{dc}{dt} = \frac{r}{V}(a - c).$$

This differential equation has the same form as that which models the cooling of a body according to Newton's law (see the module *A Hot Time*) and has the unique solution

$$c(t) = a + (c_0 - a)e^{-rt/V},$$

where c_0 is the initial concentration of solute in the vessel. Given the parameters a, c_0, V, and r, this solution of the direct problem determines the concentration of solute for all time.

This simple model suggests a number of interesting inverse problems. For example, suppose the vessel is an underground cistern into which polluted groundwater is seeping. By sampling the contents of the cistern (say, by use of a preinstalled probe), concentration measurements of the liquid in the cistern can be made at various times, and these measured concentrations might be used to try to determine the parameters of the model. The model can be generalized in various ways. For example, the case of differing inflow and outflow rates results in a model in which the volume enters as a time-dependent parameter. Inverse problems can also be posed in which the flow rates or inflow concentration are time-dependent. These and other ideas are explored in the Activities.

4.1.2 Activities

1. Exercise Groundwater containing an unknown (constant) concentration of pollutants seeps at an unknown (constant) rate into a cistern containing 1,000 gallons, and the well-stirred mixture leaks out at the same rate. Measurements show that the initial concentration of pollutants in the cistern is 1%. After one day the concentration of pollutants is 1.1%, and after two days it is 1.19%. What is the concentration of pollutants in the groundwater and at what rate is the groundwater seeping into the cistern?

2. Problem Suppose a tank contains a known constant volume of liquid that is initially free of pollutants, and that pollutants with an unknown constant concentration enter the tank at an unknown constant rate, while the well-stirred solution drains out at the same rate. Show that concentration values at two distinct positive times uniquely determine the flow rate and the inflow concentration.

3. Problem For the model in Problem 2, suppose the initial concentration of pollutants in the tank is unknown. Show that three concentration values uniquely determine the flow rate and inflow concentration.

4. Question For the model of Problem 2, suppose the volume is an unknown constant. Is it possible to determine the flow rate, the input concentration, and the volume from a series of concentration measurements? What can be determined?

5. Problem Suppose pollutants with an unknown time-dependent concentration $a(t)$ leak at an unknown constant rate r into a well containing a known volume V of liquid, and the well-mixed liquid leaves the well at the same rate. Provide a justification for the approximation

$$a(t_i) \approx \frac{V}{r} \frac{c(t_{i+1}) - c(t_{i-1})}{2h} + c(t_i), \quad i = 1, \ldots, n-1$$

of the unknown inflow concentration at times $t_i = ih$, $i = 1, \ldots, n-1$ in terms of the sampled concentrations $c(t_0), \ldots, c(t_n)$.

6. Computation Write a program based on the approximation method of the previous activity to approximate a variable inflow concentration at times $t_i = ih$, $i = 1, \ldots, n-1$, in the interval $(0, T)$, where $h = T/n$. Inputs to the program will be a well concentration function $c(t)$, a time T, a positive integer n, representing the number of time intervals desired for the approximation, and a parameter r/V. Your program should have the ability to blend random error of a given amplitude into the sampled values of the well concentration function. Test your program on the well concentration function $c(t) = te^{-t}$ using $r/V = .01$, and note the effect of errors on the computed inflow concentrations.

7. Problem Suppose that in the method of Problem 5 the well concentrations can be measured with an accuracy of ϵ, that is, instead of $c(t_i)$, only estimates c_i are available satisfying

$$|c_i - c(t_i)| \leq \epsilon.$$

If these estimates are used in the formula of Problem 5, and a step size of the form $h = C\epsilon^{1/3}$ is used, where C is a positive constant, show that the computed values of the inflow concentration, say a_i, satisfy

$$|a_i - a(t_i)| = O(\epsilon^{2/3}).$$

8. Project The aim of this project is to develop an alternative *implicit* method for approximating time-dependent inflow concentrations from well concentration measurements. For a given time $T > 0$, and positive integer n, let $h = T/n$ and $t_i = ih$, $i = 1, \ldots, n$. Let l_i be the continuous function defined on $[0, T]$ that is linear on each of the subintervals $[t_j, t_{j+1}]$ and satisfies $l_i(t_j) = 0$ for $i \neq j$ and $l_i(t_i) = 1$. Sketch the graph of l_i to see why it is sometimes called a "tent" function. The equation relating the well concentration $c(t)$ and the inflow concentration $a(t)$ is

$$\frac{V}{r}\frac{dc}{dt} = a - c,$$

where V is the constant volume and r is the constant inflow (= outflow) rate. Show that if this equation is multiplied by l_i and then integrated over $[0, T]$, the result is

$$-\frac{V}{r}\int_0^T c(t)l_i'(t)\,dt = \int_0^T a(t)l_i(t)\,dt - \int_0^T c(t)l_i(t)\,dt$$

for $i = 1, \ldots, n-1$. Given concentrations $c(t_j)$, for $j = 0, \ldots, n$, explain why

$$c(t) \approx \sum_{i=0}^{n} c(t_i)l_i(t)$$

is a reasonable approximation of the well concentration function. Assume a similar approximation for the unknown inflow concentration on $(0, T)$, that is,

$$a(t) \approx \sum_{j=1}^{n-1} a_j l_j(t),$$

where the coefficients a_j are to be determined. Substitute these approximations into the equation above to get a system of linear equations that determines the approximations a_j to the unknown inflow concentration function $a(t)$ at times $t = t_j$. Implement this scheme in software and compare the performance of this proposed method with that of the method in Computation 6.

9. Problem Suppose a solute with constant concentration a enters a well at a rate r, and the well-mixed solution leaves at a rate $\rho < r$. Let t_1 and t_2 be any two times that do not exceed the "overflow" time. Show that the initial concentration of solute in the well, the initial volume of liquid in the well, the concentration of solute in the well at time $t_1 > 0$, and the volume and concentration at a time $t_2 > t_1$ uniquely determine the inflow concentration a, the inflow rate r, and the drain rate ρ.

10. Calculation A 230-gallon tank initially contains 50 gallons of a 1% salt solution. Salt water with an unknown constant concentration flows into the tank at an unknown constant rate, while the well-stirred solution drains out of the tank at an unknown, but slower, constant rate. After $2\frac{3}{4}$ hours the concentration of salt in the tank is 2.1%, while after $9\frac{1}{2}$ hours the concentration of salt is 3.0%. The tank overflows at 20 hours. Estimate the fill rate, the drain rate, and the concentration of salt in the inflow.

11. Problem Suppose a pollutant with a known constant concentration a enters a tank containing a known volume V of liquid at an unknown time-dependent rate $r(t)$, and the well-stirred liquid drains out at the same rate. Develop a finite difference method, analogous to that of Problem 5, for approximating $r(t)$ in the interior of a given time interval $[0, T]$ in terms of measured concentration values.

12. Computation Develop software for the method of Problem 11 and test it on some models in which the rate function $r(t)$ and the concentration function $c(t)$ are known.

4.1.3 Notes and Further Reading

The direct problem for one-compartment mixing models is treated in almost all elementary differential equations texts. The multicompartment model is more challenging and requires the use of techniques from linear algebra. A particularly interesting two-compartment model involves the transfer of nutrients and wastes through the placenta; see, e.g., G. Fulford, P. Forrester, and A. Jones, *Modelling with Differential and Difference Equations*, Cambridge University Press, Cambridge, 1997.

4.2 Slip Sliding Away

Course Level:

Differential Equations

Goals:

Foster basic modeling techniques using differential equations. Develop physical intuition for resisted motion. Illustrate polar graphs in a physical context.

Mathematical Background:

Elementary differential equations (linear second-order, separable), curves in polar coordinates, parametric curves

Scientific Background:

Galileo's law of falling bodies, Newton's law of motion

Technology:

Graphics calculator

4.2.1 Introduction

Nearly everyone has heard the (perhaps apocryphal) story of Galileo testing gravity by dropping objects from the leaning tower of Pisa. With the very crude devices for time measurement available to Galileo, it would have been impossible for him to arrive at his quantitative law for falling bodies by direct observation of falling bodies—things just happen too quickly for objects in free fall. It seems that Galileo came to his law by what we would call today a "thought experiment." The law was then tested experimentally in a controlled setting by effectively slowing down the action by replacing free fall with descent down inclined planes. In the actual experiments balls were rolled down gently sloping planes, and the idealized motion of vertical free fall was deduced from these observations. In this module we consider some aspects of the motion of a point particle of unit mass that descends from a given point under the influence of a constant gravitational force. In particular, we are interested in the shape of the curve containing all positions the particle attains in a given fixed period of time while descending under the influence of gravity from a fixed point. We investigate both unresisted descent and certain types of resisted descent. The particular inverse problem that interests us is that of determining the nature of the resistance from the shape of the "equitemporal curve." This problem will be solved for only the simplest cases.

The stimulus for this module is a brief discussion in Ernst Mach's *The Science of Mechanics* in which Mach mentions a "pretty" theorem of Galileo that is little discussed in the classroom nowadays. The theorem appears in Galileo's *Two New Sciences*:

> If from the highest... point of a vertical circle any inclined planes whatever are drawn to its circumference, the times of descent through these are equal.

What Galileo is saying is that, under the influence of a constant vertical force of gravity, and neglecting resistance, the times of descent down the line segments shown in Figure 4.1 are equal. This is easy to verify by using Newton's

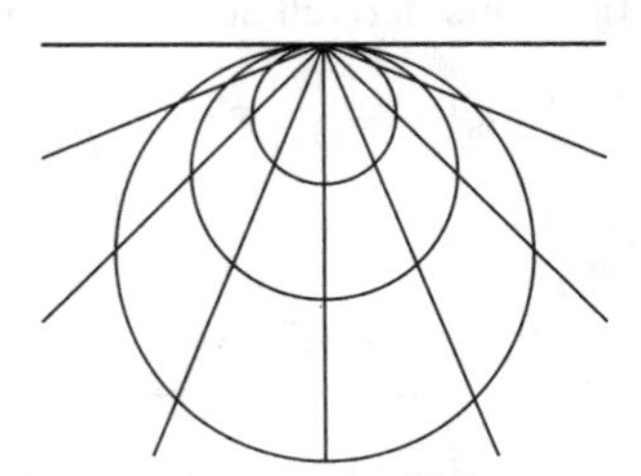

Figure 4.1: Descent from a Common Point

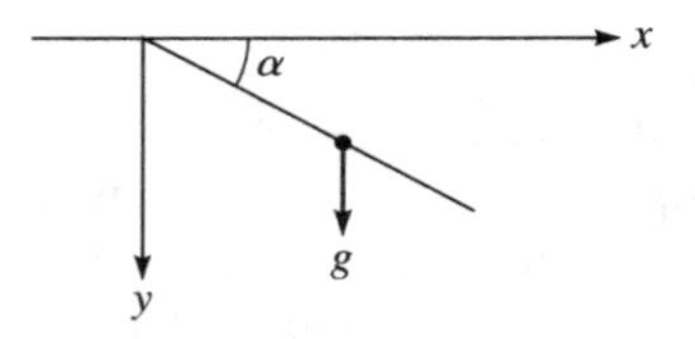

Figure 4.2: Descent Down a Ramp

laws of motion. Suppose we set up a coordinate system with the positive y-direction pointing down and consider a segment emanating from the origin and declined at an angle α to the positive horizontal axis, as in Figure 4.2.

If we let s represent the position of the particle (i.e., its distance down the ramp from the origin), then, since the component of the gravitational force in the direction of the ramp is $g\sin\alpha$, we have by Newton's law

$$\frac{d^2s}{dt^2} = g\sin\alpha, \quad s(0) = 0, \quad \frac{ds}{dt}(0) = 0$$

(recall that we assume that the particle has mass 1 and we ignore resistance). This is a very simple differential equation that may be solved immediately to give

$$s = \frac{g}{2}(\sin\alpha)t^2.$$

Note that in the special case of vertical fall ($\alpha = \pi/2$), we recover Galileo's law of falling bodies.

Suppose T is a given fixed time. The curve of points reached in time T for various declination angles α can be obtained from the equation above. We get a clearer picture of this curve if we convert to more familiar rectangular coordinates. Note that

$$x = s\cos\alpha, \quad y = s\sin\alpha;$$

therefore, using the equation for s derived above we have

$$x^2 = \frac{g}{2}T^2\sin^2\alpha\frac{g}{2}T^2\cos^2\alpha = y\frac{g}{2}T^2\cos^2\alpha = y\left(\frac{g}{2}T^2 - y\right),$$

and hence

$$x^2 + \left(y - \frac{g}{4}T^2\right)^2 = \left(\frac{g}{4}T^2\right)^2,$$

that is, the curve of positions attained in time T is a circle of diameter $(g/2)T^2$ that passes through the origin.

The direct problem of determining the form of the equitemporal curves, for certain given resistance laws, as well as the inverse problem of determining the form of the resistance law from knowledge of the shapes of the equitemporal curves, is treated in the Activities.

4.2.2 Activities

1. Problem Suppose a particle descends, under the influence of constant gravity and without resistance, on an inclined plane starting from a point on a given vertical circle and passing through the lowest point on the circle. Show that the time of descent to the lowest point of the circle is independent of the starting point.

2. Problem Establish the converse of Problem 1. Namely, the curve describing all initial positions that produce a given descent time to a given point (under constant gravity and no resistance) is a circle whose lowest point is the given point.

Henceforth, we consider particles of unit mass descending from a fixed origin under the influence of a constant gravitational acceleration g acting vertically downward, as in the figure above. But now we will consider the effect of some simple resistance laws. The curve (parameterized by α) of all points to which the particle descends in a given fixed amount of time will be called an *equitemporal curve*. We will consider both the direct problem of determining the equitemporal curves, given a resistance law, and inverse problems of determining the law of resistance from the shape of the equitemporal curves.

3. Calculation Plot the graph of the polar equation $r = D\sin\theta$ for $0 \leq \theta \leq \pi$ for various values of $D > 0$. Show that the resulting curve is a circle passing through the origin having diameter D.

4. Exercise Suppose a particle of mass 1 descends from rest at the origin along a line declined at an angle α with respect to the x-axis, as in the figure above. If the particle is resisted by a force $f(\dot{s})$ that depends only on its velocity $\dot{s}$, show that the distance s traversed satisfies $\ddot{s} = g\sin\alpha - f(\dot{s})$ (we use the Newtonian notation for derivatives with respect to time).

5. Problem Suppose that the particle in Exercise 4 experiences resistance proportional to its velocity (i.e., $f(\dot{s}) = k\dot{s}$, for some positive constant k). Show that the curve of all points reached in time t is the circle

$$s = D_k(t)\sin\alpha$$

with diameter

$$D_k(t) = \frac{g}{k}\left(t - \frac{1 - e^{-kt}}{k}\right).$$

6. Problem For fixed k, show that the diameters of the equitemporal curves in Problem 5 increase with respect to time t. What is the rate of increase? Find the limiting values of the rate of increase of the diameters as $t \to 0^+$ and as $t \to \infty$.

7. Exercise Show that as the resistance goes to zero in Problem 5, the diameters of the equitemporal circles go to that of free fall; i.e., show that

$$\lim_{k \to 0^+} D_k(t) = \frac{g}{2}t^2$$

where $D_k(t)$ is given in Problem 5.

8. Problem Suppose the resistance is a function of velocity alone, say $f(\dot{s})$, as in Exercise 4. Suppose also that the equitemporal curves are circles

$$s = D(t) \sin \alpha$$

of diameter $D(t)$. Show that

$$f(\sigma \dot{D}(t)) = \sigma f(\dot{D}(t))$$

for all $\sigma \in [0, 1]$. Conclude that the resistance is proportional to the velocity.

9. Problem Suppose the resistance is proportional to the distance descended, i.e., the resistance is of the form $f(s) = ks$ for some positive constant k. Show that the equitemporal curves are circles

$$s = D_k(t) \sin \alpha$$

of diameter

$$D_k(t) = \frac{g}{k}\left(1 - \cos\left(\sqrt{k}t\right)\right).$$

10. Exercise Show that as the resistance goes to zero in Problem 9, the diameters of the equitemporal circles go to those of free fall.

11. Problem Suppose the resistance is a function of distance along the ramp alone, i.e., the resistance has the form $f(s)$. Show that if the equitemporal curves are circles of diameter $D(t)$, then

$$f(\sigma D(t)) = \sigma f(D(t))$$

for all $\sigma \in [0, 1]$. Conclude that the resistance is proportional to s.

12. Problem Suppose the resistance is a function of distance and velocity, i.e., $f(s, \dot{s})$, and that the equitemporal curves are given by

$$s = g(1 - (1 + t)e^{-t}) \sin \alpha.$$

Find the resistance function.

13. Problem Suppose that the resistance has the form $f(\dot{s}) = k\dot{s}^2$. Show that the equitemporal curves are described by

$$s = \frac{1}{k} \ln \left[\cosh(\sqrt{kg \sin \alpha}\ t) \right].$$

14. Calculation For $k = 5$ and $t = 1, 2, 5, 10$, plot the equitemporal curves in Problem 13.

15. Problem Suppose the resistance is a function of the velocity only and the equitemporal curves are those given on Problem 13. Show that the resistance is proportional to the square of the velocity.

16. Problem Suppose the resistance function has the form $f(s, \dot{s}) = as + b\dot{s}$, for some positive constants a and b. Describe the equitemporal curves.

The remaining activities are concerned with the family of curves given parametrically by

$$x = a\theta - a \sin \theta$$

$$y = a - a \cos \theta,$$

where a is a positive constant and $0 \leq \alpha \leq \pi/2$.

17. Calculation Plot the curves corresponding to $a = .1, 5, 50, 200$.

18. Calculation Estimate a curve from the family above that passes through $(2, 3)$.

19. Problem Show that every point (x, y) with $x > 0$ and $y/x > 2/\pi$ lies on precisely one curve from the family above.

20. Problem Suppose that a unit point mass is released from rest at the origin and descends along one of the curves above (remember in our descent problems the positive y-axis is pointing downward). Find the equitemporal curves for this descent problem (i.e., for a given descent time t, find the set of points (x_a, y_a) reached in time t if the particle descends without resistance along the curve above with parameter a).

4.2.3 Notes and Further Reading

Galileo spends much of the third day of his dialogue in *Two New Sciences* on the problem of descent without resistance along straight line segments. He then speculates on the curve of quickest descent. In a scholium he erroneously states that the curve of quickest descent is the arc of a circle. In 1696, Johann Bernoulli posed the problem of finding the curve of quickest descent through two given points, the *brachistochrone problem*. The solution of this problem is a cycloid, that is, a curve from the family just studied. The problem was solved by the great mathematicians of the day, including Bernoulli, Huygens, Newton, Leibniz, and others. The brachistochrone problem led to the founding of a new branch of analysis—the calculus of variations. The history of the subject can be found in H. H. Goldstine, *The History of the Calculus of Variations from the 17th through the 19th Century*, Springer-Verlag, New York, 1980. The material in this module is an expanded version of C. W. Groetsch's, "Equitemporal curves for resisted descent" in *Mathematics and Computer Education* **31** (1997), pp. 152–157.

4.3 It's A Drag

Course Level:

Differential Equations, Elementary Analysis

Goals:

Study projectile motion with linear drag, particularly the inverse range problem.

Mathematical Background:

Linear differential equations, intermediate value theorem, mean value theorem, fixed points, Taylor's theorem

Scientific Background:

Newton's law of motion

Technology:

Graphing calculator, MATLAB or other high level numerical/graphical software

4.3.1 Introduction

In this module we revisit the inverse problem for projectiles, but now we take into account (in an admittedly naive way) the drag on the projectile due to air resistance. It is common experience that resistance increases with velocity—the faster an object moves, the more resistance it experiences. We shall develop a model in which the resistance is *proportional* to the velocity. While this is not realistic (air resistance is a very complicated phenomenon and in the most widely accepted model the resistance is assumed to be proportional to the *square* of the velocity), it is a model that can be viewed as a first approximation to resistive behavior. Newton himself studied the linear resistance model in Book II of the *Principia*. He was well aware that the linear model was not the most suitable physical model, for he wrote:

> However, that the resistance of bodies is in the ratio of velocity, is more a mathematical hypothesis than a physical one.

Nevertheless, he evidently felt that the linear model was a good warm-up before taking on the squared velocity resistance model. We will see that the linear model raises a number of interesting and challenging elementary mathematical issues.

We use the same setup as in the module *A Cheap Shot*, where trajectories in a resistanceless medium were considered. That is, we assume that the point projectile has unit mass and is launched from the origin with speed v (the *muzzle velocity*) at an angle of elevation θ with the positive x-axis. Recall that, in the absence of resistance, the range, as a function of the angle of elevation θ, is given by

$$R(\theta) = \frac{v^2}{g} \sin 2\theta,$$

where g is the acceleration due to gravity. So the direct problem of finding the range, given the angle of elevation, has a unique solution, which is given by this equation. On the other hand, the inverse problem—that of determining an angle that produces a given suboptimal range R—has precisely two solutions for each range R satisfying $0 \leq R < v^2/g$ (a fact first enunciated by Tartaglia in 1537). Furthermore, the maximum range occurs precisely at $\theta = \pi/4$ radians.

We now build a model in which the projectile experiences a resistive force that is proportional to the velocity. The constant of proportionality k is called the *resistance constant*, and if the projectile is at position (x, y) it will be subject to a resistance force (we use Newton's "dot" notation for time derivatives)

$$-k \begin{pmatrix} \dot{x} \\ \dot{y} \end{pmatrix}$$

and a gravitational force

$$\begin{pmatrix} 0 \\ -g \end{pmatrix}.$$

The equations of motion of the projectile are therefore

$$\ddot{x} = -k\dot{x}$$
$$\ddot{y} = -g - k\dot{y},$$

and the projectile satisfies the initial conditions

$$\dot{x}(0) = v\cos\theta, \quad x(0) = 0$$
$$\dot{y}(0) = v\sin\theta, \quad y(0) = 0.$$

These differential equations may be solved routinely by the familiar method for linear differential equations. For example, the first equation can be reduced to a first-order linear equation with the substitution $u = \dot{x}$, giving

$$\dot{u} + ku = 0,$$

or, equivalently,

$$0 = e^{kt}\dot{u} + ke^{kt}u = \frac{d}{dt}(e^{kt}u),$$

and therefore, taking the initial condition into account,

$$\dot{x} = u = v\cos\theta e^{-kt}.$$

A further integration, using the second initial condition, then gives

$$x = v\cos\theta(1 - e^{-kt})/k.$$

Similarly, a single integration of the differential equation for the y-coordinate function yields

$$\dot{y} = \left(v\sin\theta + \frac{g}{k}\right)e^{-kt} - \frac{g}{k},$$

and another integration (and use of the initial condition) gives

$$\begin{aligned} y &= -\left(\frac{v\sin\theta}{k} + \frac{g}{k^2}\right)e^{-kt} - \frac{g}{k}t + \frac{v\sin\theta}{k} + \frac{g}{k^2} \\ &= \left(\frac{v\sin\theta}{k} + \frac{g}{k^2}\right)(1 - e^{-kt}) - \frac{g}{k}t. \end{aligned}$$

From the equation for x, we obtain

$$1 - e^{-kt} = \frac{kx}{v\cos\theta},$$

and hence, solving for the parameter t, we find

$$t = -\frac{1}{k}\ln\left(1 - \frac{kx}{v\cos\theta}\right).$$

Substituting these results into the equation for y we arrive at

$$y = \left(\frac{v\sin\theta}{k} + \frac{g}{k^2}\right)\frac{kx}{v\cos\theta} + \frac{g}{k^2}\ln\left(1 - \frac{kx}{v\cos\theta}\right).$$

The range of the projectile is then a root x of the equation

$$x\left(\frac{k^2}{g}\tan\theta + \frac{k}{v}\sec\theta\right) + \ln\left(1 - \frac{kx}{v\cos\theta}\right) = 0,$$

or, equivalently,

$$1 - \frac{kx}{v\cos\theta} = e^{-A(\theta)x},$$

where

$$A(\theta) = \frac{k}{v}\sec\theta + \frac{k^2}{g}\tan\theta.$$

If we denote the range by $R(\theta)$, then we finally arrive at the equation that will be fundamental to the rest of our analysis:

$$R(\theta) = \frac{\cos\theta}{a}(1 - e^{-A(\theta)R(\theta)}),$$

where $A(\theta) = a\sec\theta + b\tan\theta$, $a = k/v$, and $b = k^2/g$.

In a resistanceless medium we found an *explicit* formula for the range. Now we find that when the medium offers resistance that is proportional to the velocity, the range $R(\theta)$ is given by the *implicit* relationship above. This form of relationship is particularly convenient for applying an ancient approximation method called the method of *successive approximation* or *fixed point iteration*. For a given θ this method approximates the range $R(\theta)$ by starting with an initial approximation R_0 and generating successive approximations by the recursive formula

$$R_{n+1} = \frac{\cos\theta}{a}\left(1 - e^{-A(\theta)R_n}\right).$$

In the activities below, the reader is asked to verify the convergence of this simple algorithm. The point is that, instead of computing $R(\theta)$ from an explicit formula, as can be done in a resistanceless medium, we must now compute $R(\theta)$ by a numerical algorithm. Nonetheless, the implicit expression for $R(\theta)$ given above will allow fairly extensive *analytical* studies of the direct and inverse range problems.

The MATLAB program 'shot' will compute the range $R(\theta)$ for a given value of the resistance constant k, the muzzle velocity v, and the angle of elevation θ. The program 'range' computes pairs $(\theta, R(\theta))$ for any number of equally spaced angles θ in $[0, \pi/2]$. Plotting these pairs then exposes the graph of the range function $R(\theta)$. For example, using the resistance constant $k = 1$, 'range' produced the plots in Figure 4.3, which show the range functions (not the trajectories!) for various muzzle velocities ($v = 50, 200, 500$, respectively, bottom to top). In the activities the reader is invited to use this program and several others to explore projectile motion in a linearly resisting medium.

Our primary interest is in the problem first studied seriously by Nicolo Tartaglia, the *inverse* range problem. The implicit relationship derived above can also be used to prove analytically that Tartaglia's observation, namely that to each suboptimal range there corresponds exactly two distinct angles

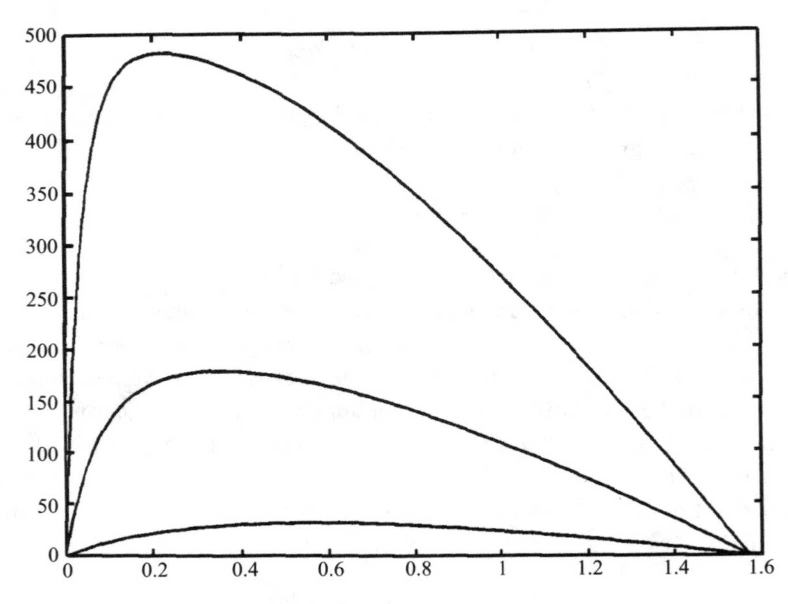

Figure 4.3: Range vs Angle of Elevation

of elevation, holds in a linearly resisting medium. Moreover, the relationship can be used to solve this inverse problem computationally, that is, calculate the angles leading to a given range. Finally, the relationship can be used to obtain a similar relationship for the optimal angle of elevation, that is, the angle that produces the maximum range. The activities include analytical and computational studies of this inverse problem.

4.3.2 Activities

1. Calculation For various constants $c > 0$, and $d > 1/c$, plot the graph of the function $f(x) = c(1 - e^{-dx})$ for $x \geq 0$. On the same screen, also plot the graph of the function $g(x) = x$. What do you observe about the intersection of these graphs?

For the next seven activities, the function f is given by

$$f(x) = c(1 - e^{-dx}),$$

where $c > 0$ and d, with $cd > 1$, are given constants. The aim of these activities is to explore the fixed points of f. We will then put this fixed-point idea to work in studying the range function of the projectile.

2. Exercise A number p is called a *fixed point* of a function f if $f(p) = p$. Show that 0 is a fixed point of the function f from Activity 1.

3. Problem Show that $f'(0) > 1$ and conclude that $f(s) > s$ for all sufficiently small positive numbers s.

4. Problem Show that $f(c) < c$. Use this and the result of the previous problem to conclude that f has a fixed point in the interval $(0, c)$.

5. Problem Show that $f''(x) < 0$ for all x, and use this to argue that f has a unique *positive* fixed point, hereafter called p.

6. Problem Show that if $f(x) > x$ for some $x > p$, then f has a fixed point q satisfying $q > p$. Conclude that if $f(x) > x$ for some x, then $x < p$, while if $f(x) < x$ for some x, then $x > p$.

7. Problem Using the inequality $e^x > 1 + x$ for $x > 0$, show that $(x + 1)(1 - e^{-x}) > x$ for $x > 0$. Use this result in turn to show that $f((cd - 1)/d) > (cd - 1)/d$. Conclude that $d(c - p) < 1$.

8. Problem Show that $0 < f'(p) < 1$.

9. Problem Let

$$F(\theta, r) = \frac{\cos\theta}{a}\left(1 - e^{-A(\theta)r}\right)$$

for $0 < \theta < \pi/2$ and $r > 0$, where $A(\theta) = a\sec\theta + b\tan\theta$.

(i) Show that F has continuous partial derivatives F_θ and F_r.

(ii) Show that $R(\theta) = F(\theta, R(\theta))$.

(iii) Use Problem 8 to show that $F_r(\theta, R(\theta)) < 1$.

10. Problem Use Problem 9 to show that R is differentiable on $(0, \pi/2)$ and continuous on $[0, \pi/2]$.

11. Problem Show that $R(\theta)$ is a solution of

$$R(\theta) = \frac{\cos\theta}{a}\left(1 - e^{-A(\theta)R(\theta)}\right)$$

if and only if $t = R(\theta)\sec\theta$ is a root of the function

$$g(t) = at - 1 + e^{-\left(at + b\sqrt{t^2 - R(\theta)^2}\right)}.$$

12. Problem Let

$$R^* = \max_{\theta\in[0,\pi/2]} R(\theta).$$

Show that if $0 \le R < R^*$, then there are exactly two angles $\theta \in [0, \pi/2]$ with $R(\theta) = R$. (Hint: If there are more than two such θ, then the function g in Problem 11 has at least three roots.)

13. Problem Consider fixed-point iteration defined by $x_{n+1} = f(x_n)$, where f is defined as in Calculation 1. Show that if $0 < x_0 < p$, where p is the unique positive fixed point of f, then $x_n < x_{n+1} < p$, while if $p < x_0$, then $p < x_{n+1} < x_n$. Conclude that fixed-point iteration converges monotonically to p for each $x_0 > 0$. Illustrate this result graphically.

14. Exercise Consider the equation for the range $R(\theta)$:

$$R(\theta) = \frac{\cos\theta}{a}\left(1 - e^{-A(\theta)R(\theta)}\right).$$

Show that for any $\theta \in (0, \pi/2)$, $R(\theta)$ is the positive fixed point of a function of the type studied in the previous activities. Conclude that fixed-point iteration may be used to compute $R(\theta)$.

15. Computation The MATLAB program 'shot' (invocation: r=shot(k,v, theta);) uses fixed-point iteration to compute the range $R(\theta)$ for given values of the resistance constant k, the muzzle velocity v, and the angle of elevation θ. For fixed values of k and v, use 'shot' to compute the range for $\theta = \pi/4$ and for various angles somewhat less than $\pi/4$. Are the results different from what you would expect to see were the medium to offer no resistance?

16. Computation Use the MATLAB program 'range' (invocation: [t,r]= range(k,v,n);) to compute and plot the graph of the range function for a fixed resistance constant and various muzzle velocities. Do the same, but this time keep the muzzle velocity fixed and vary the resistance constant.

17. Computation Activity 12 shows that Tartaglia's inverse problem has exactly two solutions θ for each suboptimal range R. Problem 11 suggests a method for calculating these angles. Namely, one finds the roots t_1, t_2 of the function $g(t)$ defined in Problem 11 and then sets $\theta_i = \sec^{-1}(t_i/R), i = 1, 2$. The MATLAB program 'theta' uses Newton's method to compute these angles, which solve the inverse problem, for given values of the resistance constant k, the muzzle velocity v, and the desired range R. The program also requires initial approximations to the two angles that may be obtained from the graphs produced by 'range'. Use 'theta' (invocation: th=theta(k,v,R,th1,th2);) to solve the inverse range problem for various values of the physical parameters. Verify your results by solving the direct problem (using the program 'shot') with the angles obtained from 'theta'.

18. Problem We now address the problem of computing the maximum range in a linearly resisting medium. If the maximum range occurs at an angle θ, then $R'(\theta) = 0$. Show that for this optimal θ we have

$$\sin\theta = (\sin\theta + c)e^{-A(\theta)R(\theta)},$$

where $c = vk/g$.

19. Problem Using the fact that $e^{-A(\theta)R(\theta)} = 1 - a\sec\theta R(\theta)$, show that the optimal angle of elevation θ satisfies

$$R(\theta) = \frac{\frac{c}{a}\cos\theta}{\sin\theta + c}.$$

20. Problem Conclude from Problem 19 that the optimal angle of elevation θ satisfies

$$A(\theta)R(\theta) = \frac{c + c^2\sin\theta}{\sin\theta + c},$$

and hence, by Problem 18,

$$s = (s + c)e^{-(c+c^2 s)/(s+c)},$$

where $s = \sin\theta$ is the sine of the angle that produces the maximum range.

21. Problem Show that the equation defining s, given in Problem 20, is equivalent to

$$x = e^{hx},$$

where $h = (1 - c^2)/(e)$ and $x = (es)/(s + c)$.

22. Problem Show that for each $h < 1/e$, the equation $x = e^{hx}$ has a unique solution $x(h) \in (0, e)$. Moreover, if $a \in (0, e)$ satisfies $e^{ha} < a$, then $x(h) < a$.

23. Computation Problem 21 suggests a way to compute the optimal angle of elevation. Solve the equation $x = e^{hx}$, then set $s = cx/e - x$ and $\theta = \sin^{-1} s$. The MATLAB program 'thopt' uses this idea (the equation is solved by Newton's method) to compute the optimal angle of elevation for given values of k and v (invocation: th=thopt(k,v);). Use 'thopt' to find the optimal angle of elevation, then use 'shot' to find the maximum range. Do this for various values of k and v.

24. Computation Write a MATLAB program to construct and plot the graph of the optimal angle of elevation versus the parameter $c = kv/g$. What do you observe?

25. Problem Suppose that j is a function having a continuous second derivative on $[1, \infty)$ and satisfying $j(1) = j'(1) = 1$. Use Taylor's theorem to show that if $j''(w) > 0$ for all $w > 1$, then $j(w) > w$ for all $w > 1$.

26. Problem Show that $s < 1/\sqrt{2}$, where s is the solution of Problem 20, if and only if $x(h) < e/(1 + \sqrt{2(1 - eh)})$, where $x(h)$ is the solution of Problem 21. Therefore, by Problem 22, $s < 1/\sqrt{2}$ will follow if

$$e^{h(e/w)} < \frac{e}{w},$$

where $w = 1 + \sqrt{2(1 - eh)}$.

27. Exercise Show that $eh = \frac{1}{2} + w - (w^2/2)$, where $w = 1 + \sqrt{2(1 - eh)}$.

28. Problem Show that for $w > 1$, $e^{e(h/w)} < e/w$ if $w < e^{(w-w^{-1})/2}$. Show that this last inequality follows from Problem 25. Conclude from Problem 26 that $s < 1/\sqrt{2}$ and hence the optimal angle of elevation in a linearly resisting medium is always less than $\pi/4$ radians.

4.3.3 Notes and Further Reading

Some of the material in this module can be found, in somewhat different form, in C. W. Groetsch, "Tartaglia's inverse problem in a resistive medium," *American Mathematical Monthly* **103** (1996), pp. 546–551, and in C. W. Groetsch and Barry Cipra, "Halley's Comment: Projectiles with linear resistance," *Mathematics Magazine* **70** (1997), pp. 273–280.

As mentioned in the introduction, Tartaglia held that the 45° angle produces the maximum range. According to Stillman Drake (*Galileo Studies*, University of Michigan Press, 1970, p. 26), the Veronese gunners in 1531 felt that the shot with maximum range occurs at some angle less than 45°. Of course, if air resistance is neglected, Tartaglia was right. Problem 28 shows that, in the oversimplified model in which air resistance is proportional to velocity, the Veronese gunners were right.

4.4 Ups and Downs

Course Level:

Differential Equations

Goals:

Identify constant parameters in a one-dimensional dynamical system.

Mathematical Background:

Linear constant coefficient differential equations; finite difference approximation to the derivative

Scientific Background:

Newton's law of motion; Hooke's Law

Technology:

MATLAB or other high-level numerical software

4.4.1 Introduction

Attach an object to a stiff vertical spring, push or pull the object from its equilibrium position, let it go, and watch what happens. The object typically oscillates up and down, the amplitude of the oscillations decaying with time. The amplitude, frequency, and decay rate of the oscillations depend upon the initial conditions and certain physical parameters of the system: the mass of the object, the stiffness of the spring, and the drag of the environment on the object. The *direct* problem of vibration analysis consists of determining the state of the system, that is, the position of the object relative to the equilibrium position, as a function of time, given the physical parameters. On the other hand, the *inverse* problem involves the determination of these important physical parameters from observations of the state of the system.

We will restrict our discussion to the simplest one-dimensional model, illustrated in Figure 4.4. The displacement of the object below the equilibrium position is denoted by y, and we assume that the only forces acting on the object are those exerted by the spring and viscous drag (an automobile shock absorber is a helpful mental image). The spring force is modeled by *Hooke's Law*, which states that the force acting to restore the object to its equilibrium position is proportional to the extension (or compression) of the spring. The positive constant of proportionality, k, is called the *stiffness* of the spring. The object experiences a drag force acting in the direction opposite to its motion. We take this drag to be proportional to the object's speed. The constant of proportionality, c, will be called the *damping coefficient*.

Newton's law of motion gives a mathematical description of the state:

$$m\ddot{y} = -c\dot{y} - ky,$$

where the dots indicate derivatives with respect to time. This is a linear, constant-coefficient, homogeneous differential equation, and the familiar ansatz $y = e^{rt}$ provides solutions. The nature of the solutions is determined by the

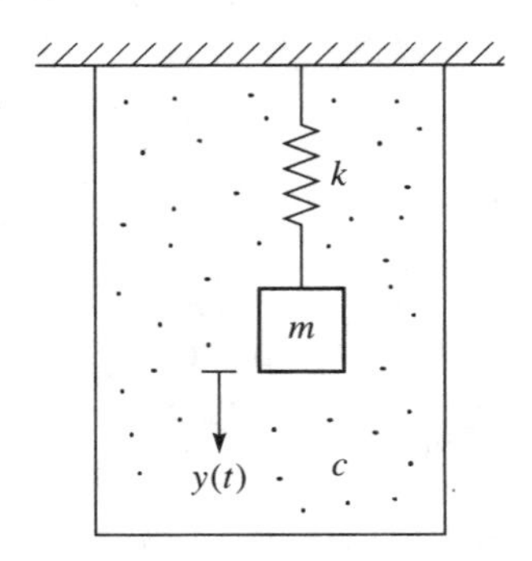

Figure 4.4: A One-dimensional Damped Vibrating System

roots of the characteristic equation

$$mr^2 + cr + k = 0.$$

Oscillatory solutions result if the characteristic equation has complex roots, that is, if $c < 2\sqrt{mk}$. In this case of a relatively small damping coefficient, we say that the system is *underdamped*. The system is *critically damped* if $c = 2\sqrt{mk}$. In this case, the solution of the equation of motion is the product of a linear function with a decaying exponential function. Therefore, there is at most one time (the root of the linear factor) at which the object is in the equilibrium position. Finally, if $c > 2\sqrt{mk}$, the system is *overdamped* and the solution is a linear combination of independent exponential functions, and hence the object released from rest from a nonequilibrium position cannot pass through the equilibrium position in positive time (see Problem 1). Knowledge of the nature of the solution of the direct problem, when combined with appropriate characteristics of the state, can sometimes be used to solve the inverse problem of determining the coefficients (see Problems 2–4 and Exercise 5).

It is clear that no degree of knowledge of the state is sufficient to determine all three parameters in the unforced system

$$m\ddot{y} + c\dot{y} + ky = 0.$$

The best one can hope for is to determine ratios, such as c/m and k/m (or, equivalently, determine the damping coefficient and stiffness, assuming a unit mass). In the remainder of this introduction we will therefore limit our attention to unit–mass systems of the form

$$\ddot{y} + c\dot{y} + ky = 0.$$

If the state is observed at certain times $t_i, i = 0, 1, \ldots, n + 1$, where $t_{i+1} - t_i = h > 0$, one can approximate the derivatives by the well-known finite difference quotients

$$\ddot{y}(t_i) \approx \frac{y(t_{i+1}) - 2y(t_i) + y(t_{i-1})}{h^2}$$

$$\dot{y}(t_i) \approx \frac{y(t_{i+1}) - y(t_{i-1})}{2h}.$$

Substituting these approximations into the equation of motion results in the equations

$$\frac{h}{2}(y(t_{i+1}) - y(t_{i-1}))c + h^2 y(t_i)k = -y(t_{i-1}) + 2y(t_i) - y(t_{i+1}),$$

$$i = 1, \ldots, n - 1$$

for approximating c and k. Of course, we have only two unknowns to approximate, but the number of equations above is limited only by our ability to sample the state. This system of $n-1$ equations in two unknowns is overdetermined (if $n > 3$) and generally has no solution in the strict sense. There is, however, a standard technique for obtaining a *least-squares* solution, that is, a pair of numbers (c, k) that minimizes the sum of the squares of the deviations in the equations. MATLAB has the built-in ability to calculate least-squares solutions of overdetermined systems. This feature is used in the program 'coeff', which allows the user to experiment with the finite-difference least-squares method for approximating the stiffness and damping coefficient. The program accepts a function y satisfying the equation of motion for an unforced unit mass, a final time T, a positive integer n, and a parameter 'ep'. It approximates c and k by using the state values $y(ih), i = 0, \ldots, n; h = T/n$, which have been corrupted with uniform relative error of amplitude at most ep. Varying this error-level parameter allows the user to simulate the parameter recovery process with noisy data and observe the stability (or lack thereof) of the parameter identification problem.

4.4.2 Activities

1. Problem Suppose m, k, and c are positive constants. Find the general solution of the differential equation $m\ddot{y} + c\dot{y} + ky = 0$ for each of the three cases $c <, =$, and $> 2\sqrt{mk}$, and verify the statements made in the introduction. Namely, if the system is underdamped, then y is the product of a decaying exponential with a periodic function; if the system is critically damped, then y has at most one positive root; and if the system is overdamped and $y'(0) = 0 \neq y(0)$, then y has no positive root.

2. Problem An underdamped system with a 1-kilogram mass has a static deflection of .2 meters and is observed to cross the equilibrium position 9 times in 3 seconds. Find the stiffness and damping coefficient.

3. Problem A 1-kilogram object extends a spring by .21 meters. The object is disturbed and executes an underdamped motion. After 25 cycles of its motion, the amplitude has decreased by 90 percent. Find the stiffness and damping coefficient.

4. Problem Suppose a 1-kilogram mass in a critically damped system crosses the equilibrium position at a time $t_1 > 0$ and "turns around" at a time $t_2 > t_1$ (i.e., $\dot{y}(t_2) = 0$). Show that the difference $t_2 - t_1$ between the crossover time and the turnaround time uniquely determines both the stiffness and the damping coefficient.

5. Exercise A critically damped system with a 1-kilogram mass is observed to turn around 10 seconds after crossing the equilibrium position. Find the stiffness and damping coefficient.

6. Computation Using $y = (1 - .9t)e^{-.1t}$ as a test case, compute the stiffness and damping coefficient using the program 'coeff' (invocation: [x]=coeff('y',T,N,ep);). Use various times T and numbers of subintervals N. Compare the computed coefficients with the exact coefficients. Begin with $ep = 0$ and then run some simulations with $ep = .001, .01$, etc., and observe the effect of noise on the computed coefficients. The approximate damping coefficient and stiffness are returned in the first and second components, respectively, of the vector **x**.

7. Computation Repeat the previous computation using the state function

$$y = e^{-.032t} \sin 2.18t.$$

8. Computation Repeat the previous computation using the state function $y = e^{-2t} + 2e^{-t}$.

9. Exercise Explain how the program 'coeff' could be used to approximate the mass and stiffness, given the damping coefficient. Explain how it may be used to approximate the mass and damping coefficient, given the stiffness.

10. Computation Given the following observations of the state of a one-dimensional unforced system, use the program 'coeff1' (invocation: [x]=coeff1(y,N,delt);) to estimate the ratios k/m and c/m.

t	y
0.70	0.348
0.91	0.366
1.12	0.365
1.33	0.351
1.54	0.330
1.75	0.304
1.96	0.276

11. Problem Consider the equation of motion for the unforced system

$$\ddot{y} + c\dot{y} + ky = 0, \quad 0 < t < T.$$

For a given positive integer n, let $t_i = iT/n$, $i = 0, \ldots, n$. Suppose ϕ_i is the function that is linear on each of the intervals $[t_{i-1}, t_i]$ for $i = 1, \ldots, n$, and satisfies $\phi_i(t_j) = 0$ for $i \neq j$ and $\phi_i(t_i) = 1$. (To picture ϕ_i, just "connect the dots.") Multiply both sides of the equation of motion by $\phi_i(t)$ and integrate from 0 to T. Show that the following system results:

$$-\int_0^T \dot{y}(t)\dot{\phi}_i(t)\,dt - \int_0^T y(t)\dot{\phi}_i(t)\,dtc + \int_0^T y(t)\phi_i(t)\,dtk = 0,$$

for $i = 1, \ldots, n-1$. Suppose now that the approximations

$$y(t) \approx \sum_1^{n-1} y(t_j)\phi_j(t)$$

and

$$\dot{y}(t) \approx \sum_1^{n-1} y(t_j)\dot{\phi}_j(t)$$

are made in the integrals above. Show that the resulting system of $n-1$ equations is

$$\begin{aligned}\frac{h}{2}(y(t_{i+1}) - y(t_{i-1}))c + \frac{h^2}{6}(y(t_{i-1}) + 4y(t_i) + t(t_{i+1}))k \\ = -y(t_{i-1}) + 2y(t_i) - y(t_{i+1}),\end{aligned}$$

for $i = 1, \ldots, n-1$. Compare this overdetermined "finite element" system for determining c and k with the system of equations based on finite differences developed in the introduction.

12. Computation Modify the program 'coeff' to compute least-squares estimates of c and k using the finite-element system of the previous activity, instead of the finite-difference equations used in the original program. Rerun Activities 6, 7, and 8 using the modified program. When error is present in the data (i.e., when $ep > 0$), does one technique (finite element or finite difference) appear to have an edge? Can you account for this?

13. Problem Consider the forced system

$$m\ddot{y} + c\dot{y} + ky = f,$$

where $f(t)$ is an external force acting on the system at time t. Develop a method based on finite differences for estimating the three parameters m, c, and k using observations of the state, $y(t_i)$, and external force, $f(t_i)$, at times $t_i = t_0 + ih$, $i = 0, \ldots, n$.

14. Computation Modify the program 'coeff1' to estimate the parameters m, c, and k given the state and external force at discrete times. Use your program to estimate these parameters given the following data:

t	y	f
0.3	0.857	0.295
0.5	0.784	0.486
0.7	0.683	0.688
0.9	0.566	0.879
1.1	0.445	1.090
1.3	0.331	1.280

Perturb the data slightly and rerun your program; note the instability of the inverse problem.

4.4.3 Notes and Further Reading

A Laplace transform approach to coefficient identification problems of the type studied in this module can be found in Bellman, Kalaba, and Lockett's *Numerical Inversion of the Laplace Transform*, American Elsevier, New York, 1966.

4.5 A Hot Time

Course Level:

Differential Equations

Goal:

Treat some elementary inverse problems in heat flow related to identification of parameters.

Mathematical Background:

Linear differential equations, partial derivatives

Scientific Background:

Newton's law of cooling, Fourier's law of heat flow

Technology:

MATLAB or other high-level numerical software

4.5.1 Introduction

The surface temperature of a body that is not in thermal equilibrium with its surroundings changes in time. If the body is warmer than its environment, it cools as heat "flows" from the body into the environment. The simplest model of this phenomenon, *Newton's law of cooling*, holds that the rate at which the surface temperature changes in time is proportional to the difference between the ambient and surface temperatures. If the surface temperature at time t is $u(t)$, and the ambient temperature is a constant A, then by Newton's law of cooling

$$\frac{du}{dt} = \alpha(A - u),$$

where α, the heat transfer coefficient, is a positive constant. This is a classical exponential decay model, and the *direct* problem of determining the surface temperature has the unique solution

$$u(t) = A + (u(0) - A)e^{-\alpha t}.$$

This solution depends on three parameters: the ambient temperature A, the initial temperature $u(0)$, and the heat transfer coefficient α. Of course, observation of the surface temperature at appropriate times allows the solution of the *inverse* problem of determining the parameters A, $u(0)$, and α (see Exercises 2–3).

Newton's law of cooling is purely a *surface* principle; it involves a *boundary* condition, and it leads to a surface temperature that depends only on time. In extended bodies, interior temperatures typically vary not only in time but also from place to place. For example, the handle of a skillet is usually a bit cooler than the pan. The physical principles that govern internal temperatures were first explained by Joseph Fourier (1768–1830) at the beginning of the nineteenth century. Fourier's analysis hinged on the relationship between heat and temperature, and on the principle of conservation of energy.

Heat is a form of energy: The heat content of a body is a measure of the *total* kinetic energy of the molecules of the body. Temperature, as gauged by a test body (a *thermometer*), is related to the *average* kinetic energy of the molecules of a body. The heat content of a body depends not only on its temperature, but also on its mass—a 5-kilogram ball of iron at a given temperature has five times the thermal energy of a 1-kilogram ball of iron at the same temperature.

Heat is also related to the specific type of material. A 1-kilogram ball of cotton at a given temperature has less thermal energy than a 1-kilogram ball of lead at the same temperature. These ideas are bound together by the relationship

$$Q = cmu,$$

where u is the (uniform) temperature of a body, m is its mass, c is a material-dependent parameter called the specific heat of the substance, and Q is the heat content. Typical units are calories for Q, degrees Celcius for u, grams for m, and hence calories per gram-degree for c.

We limit our discussion of internal temperatures to a body with the simplest geometry—a bar of unit length and unit cross-sectional area, which we imagine to extend along the unit interval of the x-axis. Suppose the mass density and specific heat of the material of which the bar is made are ρ and c, respectively. We assume that the lateral surface of the bar is insulated so that the spatial dependence of the temperature is a function of the single variable x. The temperature of the point of the bar at position $x \in [0, 1]$ and at time $t \geq 0$ is then a function $u(x, t)$. Consider a thin slice of the bar extending over the interval $[x, x + \Delta x]$. The heat content of this slice is then about

$$c\rho u \Delta x,$$

and the rate of change of this thermal energy with respect to time is approximately

$$\frac{\partial(c\rho u)}{\partial t}\Delta x.$$

Respect for the principle of conservation of energy demands that this quantity equal the net rate of flow of thermal energy into the slice, plus the rate, if any, at which heat is produced within the slice. If the rate at which heat is produced (at position x and time t) per unit volume is denoted by $f = f(x, t)$, then the rate at which heat is produced within the slice is approximately $f\Delta x$ (remember that we assume a unit cross-sectional area).

Heat may flow into (or out of) the slice only through the left face at x or the right face at $x + \Delta x$. Fourier's law, the final ingredient in the model, states that the rate of flow of heat through a face is proportional to the negative *temperature gradient*, $-\partial u/\partial x$, at the face (the reason for the negative sign is that heat flows from hot to cold). The net flow of heat across the surface into the slice $[x, x + \Delta x]$ is therefore

$$k\frac{\partial u}{\partial x}\bigg|_{x+\Delta x} - k\frac{\partial u}{\partial x}\bigg|_{x},$$

where the proportionality constant, k, is called the thermal conductivity. Adding this to the rate at which thermal energy is generated internally, we find that the net rate of change of thermal energy within the slice is approximately

$$k\frac{\partial u}{\partial x}\bigg|_{x+\Delta x} - k\frac{\partial u}{\partial x}\bigg|_{x} + f\Delta x.$$

By the conservation of energy principle, this should match the rate calculated previously, that is,

$$\frac{\partial(c\rho u)}{\partial t}\Delta x \approx k\frac{\partial u}{\partial x}\bigg|_{x+\Delta x} - k\frac{\partial u}{\partial x}\bigg|_{x} + f\Delta x.$$

The precise model results when the interval $[x, x+\Delta x]$ is shrunk to the point x:

$$\frac{\partial(c\rho u)}{\partial t} = \lim_{\Delta x\to 0}\frac{k\frac{\partial u}{\partial x}\big|_{x+\Delta x} - k\frac{\partial u}{\partial x}\big|_{x}}{\Delta x} + f$$

or

$$\frac{\partial(c\rho u)}{\partial t} = \frac{\partial}{\partial x}\left(k\frac{\partial u}{\partial x}\right) + f.$$

This is Fourier's celebrated *heat equation.*

The *direct* problem for the heat equation consists of finding the temperature $u(x,t)$ for all positions $x \in (0,1)$ and times $t > 0$, given certain boundary conditions, say the temperatures of the endpoints $u(0,t)$ and $u(1,t)$, an initial temperature distribution $u(x,0)$, and values of the parameters c, ρ, and k. These parameters are, in general, functions of space, time, and temperature.

In the Activities, we treat some relatively simple *inverse* problems involving the identification and estimation of "distributed" parameters in the heat model. Specifically, we consider the problem of determining the time-dependent parameter $a(t)$ in the problem

$$\frac{\partial u}{\partial t} = a(t)\frac{\partial^2 u}{\partial x^2}, \quad 0 < x < 1, \quad t > 0$$

$$u(0,t) = u(1,t) = 0$$

$$u(x,0) = \sin \pi x$$

from observations of the temperature history $h(t) = u(.5,t)$ of the midpoint of the bar.

We also propose a method for estimating the function $b(x)$ in the problem

$$b(x)\frac{\partial u}{\partial t} = \frac{\partial^2 u}{\partial x^2}, \quad 0 < x < 1, \quad t > 0$$
$$u(0,t) = u(1,t) = 0$$
$$u(x,0) = g(x)$$

from observations of u.

Finally, we study some aspects of identifying the distributed parameter $k(x)$ in the *steady state* (i.e., $\partial u/\partial t = 0$ for all x) heat distribution problem

$$\frac{d}{dx}\left(k(x)\frac{du}{dx}\right) = -f(x), \quad 0 < x < 1.$$

4.5.2 Activities

1. Exercise Suppose u is the surface temperature of a body that cools, according to Newton's law, in an environment with constant temperature $A < u(0)$. Show that $u(t)$ is a strictly decreasing function whose graph is concave-up and has $u = A$ as a horizontal asymptote.

2. Exercise Measured surface temperatures of a body that cools according to Newton's law are given at various times in the following table:

t(min.)	u(°F)
5	72
10	62
15	54

Find the ambient temperature, the initial surface temperature, and the heat transfer coefficient.

3. Problem Show that observations of the surface temperature u of a body that cools according to Newton's law at three times $t_1 < t_2 < t_3$ uniquely determine the parameters $A, u(0)$, and α.

4. Problem Show that if a body cools according to Newton's law, then for any sequence of times $t_1 < t_2 < t_3 < \cdots$ that forms an arithmetic progession, the sequence of temperature differences $A - u(t_k)$ forms a geometric progression.

5. Exercise Suppose $a(t)$ is a positive continuous function for $t \geq 0$. Give physical interpretations for the following conditions on a function $u = u(x,t)$:

$$\frac{\partial u}{\partial t} = a(t)\frac{\partial^2 u}{\partial x^2}, \quad 0 < x < 1, \quad t > 0$$
$$u(0,t) = u(1,t) = 0$$
$$u(x,0) = \sin \pi x, \quad 0 < x < 1.$$

Show that the function

$$u(x,t) = e^{-\pi^2 \int_0^t a(\tau)d\tau} \sin \pi x$$

satisfies these conditions.

6. Problem Suppose $a(t) > 0$ and $w(x,t)$ satisfies

$$\frac{\partial w}{\partial t} = a(t)\frac{\partial^2 w}{\partial x^2}, \quad 0 < x < 1, \quad t > 0$$
$$w(0,t) = w(1,t) = 0$$
$$w(x,0) = 0, \quad 0 < x < 1.$$

Show that $w(x,t) = 0$ for $0 < x < 1, t > 0$. (Hint: Let $J(t) = \int_0^1 w(x,t)^2\,dx$. Use the given conditions to show that $J(0) = 0$ and $J'(t) \leq 0$ for $t \geq 0$.)

7. Problem Use the previous problem to show that for given functions $a(t) > 0$ and $g(x)$, the problem

$$\frac{\partial u}{\partial t} = a(t)\frac{\partial^2 u}{\partial x^2}, \quad 0 < x < 1, \quad t > 0$$
$$u(0,t) = u(1,t) = 0$$
$$u(x,0) = g(x), \quad 0 < x < 1$$

has at most one solution. Therefore, the solution given in Exercise 5 is unique.

8. Problem Let $h(t) = u(.5,t), (t > 0)$ be the temperature history of the midpoint of the bar modeled in Exercise 5. Show that the variable coefficient $a(t)$ is determined by h via the formula

$$a(t) = -\frac{h'(t)}{\pi^2 h(t)}.$$

9. Problem Show that $a(t) = 1$ in Exercise 5 if and only if $h(t) = u(.5,t) = e^{-\pi^2 t}$. Let

$$d_n(t) = \begin{cases} 0, & 0 \le t \le 1 \\ -2n^5(t-1)^3 + 3n^2(t-1)^2, & 1 < t \le 1 + 1/n^2 \\ 1/n, & 1 + 1/n^2 < t, \end{cases}$$

and let $h_n(t) = h(t) + d_n(t)$, where $h(t) = e^{-\pi^2 t}$. Show that

$$|h(t) - h_n(t)| \le 1/n, \quad (t \ge 0),$$

and hence $h_n(t) \to h(t)$ as $n \to \infty$. Let $a_n(t)$ be the coefficient determined from $h_n(t)$ via Problem 8. Show that $a_n(1 + (1/2n^2)) \to \infty$ as $n \to \infty$. What does this say about the stability of the coefficient identification problem?

10. Exercise Give physical interpretations for the following conditions on a function $u = u(x,t)$:

$$b(x)\frac{\partial u}{\partial t} = \frac{\partial^2 u}{\partial x^2}, \quad 0 < x < 1, \quad t > 0$$

$$u(0,t) = u(1,t) = 0$$

$$u(x,0) = g(x), \quad 0 < x < 1.$$

11. Exercise

(a) Show that if $u_1(x,t) = e^{-\pi^2 t/4} \sin \pi x$, and $g(x) = \sin \pi x$, then the conditions of Exercise 10 are satisfied for $b(x) = 4$.

(b) Show that if $u_2(x,t) = x(1-x)e^{-t}$ and $g(x) = x(1-x)$, then the conditions of Exercise 10 are satisfied for $b(x) = 2/(x(1-x))$.

12. Problem Let $b(x)$ be the coefficient in Exercise 10 corresponding to the temperature distribution $u(x,t) = e^{-\pi^2 t} \sin \pi x$, and let $b_n(x)$ be the coefficient corresponding to

$$u_n(x,t) = u(x,t) + \frac{1}{n} e^{-\pi^2 t} \sin n\pi x,$$

where n is a positive integer. Show that

$$|u_n(x,t) - u(x,t)| \le \frac{1}{n}, \quad x \in [0,1], \quad t \ge 0,$$

but $b_n(x) \not\to b(x)$ as $n \to \infty$ for any $x \in (0,1)$.

13. Problem Let $v(x) = \int_0^\infty e^{-t} u(x,t)\,dt$ where u is a solution of the problem in Exercise 10. Show that

$$\int_0^\infty e^{-t} \frac{\partial u}{\partial t}(x,t)\,dt = v(x) - g(x)$$

and that

$$v''(x) = b(x)(v(x) - g(x)), \quad 0 < x < 1,$$

and $v(0) = v(1) = 0$.

14. Problem For a given positive integer n, let $h = 1/n$ and $x_i = i/n$, for $i = 0, \ldots, n$. Explain why the scheme

$$b(x_i) \approx \frac{v((i+1)h) - 2v(ih) + v((i-1)h)}{(v(ih) - g(ih))h^2}$$

for $i = 1, \ldots, n-1$ is a plausible approximation method for the coefficient $b(x)$ in Exercise 10 ($v(x)$ is the function defined in Problem 13).

15. Computation The Gauss–Laguerre approximation is a well-known method for estimating integrals of the form $\int_0^\infty e^{-t} f(t)\, dt$. It can therefore be used to approximate values of the function v defined in Problem 13. The routine 'temps' (invocation: [U,g]=temps(u,n,ep);) accepts a temperature function $u(x, t)$ and a positive integer n and generates discrete temperature values $u(x_i, t_i)$ at $x_i = i/n$, $i = 1, \ldots, n-1$ and at the five Gauss–Laguerre nodes $t_1, \ldots, t_5$. These numbers are returned in an $(n-1) \times 5$ matrix U. A user-supplied parameter 'ep' allows mixing of random noise of relative error amplitude ep into the discrete temperature values (set $ep = 0$ for no noise). The entries of U can then be used as a possibly noisy sample of the temperature distribution $u(x, t)$. A discretized, possibly noisy version of the initial temperature distribution is returned by 'temps' in the vector g. The program 'coeff2' (invocation: [b,x]=coeff2(U,g,n);) accepts the temperature samples in U and g supplied by 'temps' and uses the method of Problem 14 to estimate the distributed parameter $b(x)$. This approximation may then be graphed with the command 'plot(x,b)'. Use these programs with the temperature function $u_1(x, t)$ of Exercise 11 to estimate the corresponding coefficient $b(x) = 4$. Use $n = 10, 50, 100$ and plot the computed coefficient, comparing it with the true coefficient. Repeat the procedure with noise levels $ep = .001, .01, .1$ and comment on the stability of the coefficient reconstruction problem.

16. Computation Repeat Computation 15 using the function $u_2(x, t)$ of Exercise 11.

17. Problem Show that a differentiable coefficient $k(x)$ satisfying

$$-\frac{d}{dx}\left(k(x)\frac{du}{dx}\right) = f(x), \quad 0 < x < 1$$

$$-k(0)u'(0) = A, \quad -k(1)u'(1) = B$$

exists only if

$$\int_0^1 f(s)ds = A - B.$$

18. Problem Suppose f satisfies the necessary condition of Problem 17 and that $u'(x) \neq 0$ for $x \in [0,1]$. Show that the unique coefficient satisfying the conditions of Problem 17 is given by

$$k(x) = -\frac{1}{u'(x)}\left(A + \int_0^x f(s)\,ds\right).$$

19. Problem Let $f(x) = -4x + 2$, for $x \in [0, .5)$ and $f(x) = 0$ for $x \in [.5, 1]$. Suppose that $u(x) = x^2 - x + 5/4$ for $x \in [0, .5)$ and $u(x) = 1$ for $x \in [.5, 1]$. Show that if d is any function which is differentiable on $[.5, 1]$ and satisfies $d(.5) = 0$, $d'(.5) = 1$, then

$$k(x) = \begin{cases} x - .5, & 0 \leq x \leq .5 \\ d(x), & .5 \leq x \leq 1 \end{cases}$$

satisfies

$$-\frac{d}{dx}\left(k(x)\frac{du}{dx}\right) = f(x), \quad 0 < x < 1$$

$$k(0)u'(0) = k(1)u'(1) = 0.$$

Therefore, the problem of identifying the coefficient $k(x)$ has, in this case, infinitely many solutions.

20. Problem Suppose $u(x) = x$. Show that $k(x) = x$ satisfies

$$-\frac{d}{dx}\left(k(x)\frac{du}{dx}\right) = -1.$$

For a given number $\epsilon > 0$, show that

$$k_\epsilon(x) = \frac{\epsilon x}{\epsilon + \cos(x/\epsilon^2)}$$

satisfies the same equation with u replaced by

$$u_\epsilon(x) = \epsilon \sin(x/\epsilon^2) + x.$$

Show that for all $x \in [0, 1]$, $|u_\epsilon(x) - u(x)| \to 0$ as $\epsilon \to 0$, but $k_\epsilon(x) \not\to k(x)$ as $\epsilon \to 0$ for any $x \in (0, 1]$.

4.5.3 Notes and Further Reading

Newton's law of cooling is traced to an anonymous note, in Latin, published in the *Philosophical Transactions of the Royal Society of London* in 1701. The note was translated into English, identified as the work of Newton, and published as an appendix to Roger Cotes' *Hydrostatical and Pneumatical Lectures* (third edition, edited by Robert Smith, London, 1775). In this note, "A scale of degrees of heat," Newton proposed a temperature scale for a number of thermal events, as measured by the expansion of linseed oil in a glass "thermometer" tube. For example, Newton assigns a temperature of 12 degrees to "the greatest heat which a thermometer can acquire in contact with a human body: the heat of a bird hatching her eggs is much the same." In describing the cooling of iron, Newton remarks that by his scale "if the time of its cooling be divided into equal parts, the corresponding heats... will decrease in a geometrical progression, and therefore may easily be found by a table of logarithms." So it appears that Newton's law of cooling was not originally stated in terms of derivatives (fluxions), but rather in terms of the resulting exponential model (see Problem 4).

A brief biography of Fourier and a complete annotated translation of his 1807 memoir *Sur la propagation de la Chaleur* can be found in Grattan-Guiness' book *Joseph Fourier 1768–1830*, MIT Press, Cambridge, 1972. The most authoritative reference on the heat equation in a single space dimension is J.R. Cannon's *The One-Dimensional Heat Equation*, Addison-Wesley, Menlo Park, 1984. The material on the determination of the time-dependent parameter $a(t)$ in Exercise 5 and Problems 6–8 is taken from Cannon's book. The treatment of the determination of the space-dependent parameter $b(x)$ in Exercises 10–11 and Problems 12–15 was inspired by a discussion in Bellman, Kalaba, and Lockett's *Numerical Inversion of the Laplace Transform*, American Elsevier, New York, 1966.

4.6 Weird Weirs

Course Level:

Differential Equations

Goal:

Use Laplace transforms and other computational tools to study an elementary inverse problem in hydraulics.

Mathematical Background:

Laplace transforms, convolution theorem

Scientific Background:

Torricelli's Law

Technology:

MATLAB or other high-level numerical software

4.6.1 Introduction

A particularly simple inverse problem involving an integral equation occurs in irrigation theory. Consider an elevated irrigation canal in which the depth of the water is h. The wall of the canal is fitted with a weir notch that is symmetric with respect to a central vertical axis. A sluice gate in the notch may be removed to allow water to flow into a field. The canal is so large in comparison to the weir notch that the water level in the canal does not drop significantly when the gate is removed. Clearly the rate at which the water flows through the notch depends on the depth h of water in the canal and the shape of the notch. The form of the notch is specified by a *shape* function $x = f(y)$, as illustrated in Figure 4.5.

Torricelli's Law (see the module *A Little Squirt*) provides the relationship between the notch shape and the flow rate. Consider a thin horizontal slab of water of thickness Δy positioned at height y in the notch. By Torricelli's Law the velocity of this slab as it leaves the notch is $\sqrt{2g(h-y)}$. The cross-sectional area of the slab is $2f(y)\Delta y$, and hence the volume ΔV of water in the slab crossing the notch in a small time Δt is

$$\Delta V \approx 2f(y)\sqrt{2g(h-y)}\Delta y \Delta t.$$

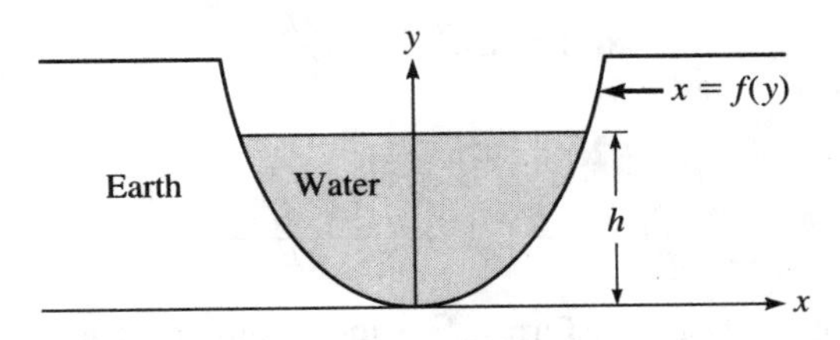

Figure 4.5: A Weir Notch

The *flow rate* function $r(h)$ giving the rate at which the volume of water exits the notch is then obtained by summing $\Delta V/\Delta t$ over all horizontal slabs and invoking the usual limit argument:

$$r(h) = 2\int_0^h \sqrt{2g(h-y)}f(y)dy.$$

We will use the figure $g = 32$ ft/sec throughout, so the basic equation relating notch shape and flow rate is

$$r(h) = 16\int_0^h \sqrt{h-y}f(y)\,dy.$$

The *direct* problem of determining the flow rate function r from the notch shape function f therefore amounts to a fairly straightforward integration. The formal solution of the *inverse* problem of determining the notch shape f from the flow rate r is also simple due to the special nature of the model equation. Namely, r is the convolution product of the function $k(u) = 16\sqrt{u}$ with the notch shape f. The convolution theorem for Laplace transforms then implies

$$R(s) = K(s)F(s),$$

where the uppercase letters indicate the Laplace transform of the function symbolized by the corresponding lowercase letter, for example,

$$R(s) = \mathcal{L}\{r\} = \int_0^\infty e^{-sh}r(h)\,dh.$$

The formal solution of the inverse problem is therefore provided by the inverse Laplace transform:

$$f = \mathcal{L}^{-1}\{R/K\}.$$

For example, if we wish to design a weir notch that has the flow rate function $r(h) = h^{3/2}$, then, since

$$R(s) = \mathcal{L}\{r\} = \frac{\Gamma(5/2)}{s^{5/2}}$$

and

$$K(s) = \mathcal{L}\{k\} = \frac{16\Gamma(3/2)}{s^{3/2}}$$

(see any table of Laplace transforms; Γ is the gamma function), we need only form a notch in the shape

$$f(y) = \mathcal{L}^{-1}\{R/K\} = \frac{1}{16}\mathcal{L}^{-1}\left\{\frac{5}{2s}\right\} = \frac{5}{32},$$

that is, the notch is a simple rectangular cutout.

4.6.2 Activities

1. Exercise Find the flow function $r(h)$ generated by the weir notch shape

$$f(y) = \begin{cases} 0 & y < 1/2 \\ 1 & y \geq 1/2 \end{cases}.$$

2. Problem Find the flow function $r(h)$ generated by the weir notch shape $f(y) = \sqrt{y}$.

3. Problem Use the method of Laplace transforms to find the weir notch shape that accounts for the flow rate function $r(h) = h^2$.

4. Problem Use the method of Laplace transforms to find the weir notch shape that accounts for the flow rate function $r(h) = h^{5/2}$.

5. Problem Find a weir notch shape that results in a "proportional" flow rate function of the form $r(h) = ch$. Is this exact shape physically realizable?

6. Problem Show that, if f is bounded, then the flow rate function satisfies $r(0) = r'(0) = 0$.

7. Problem Let $F(s)$ and $R(s)$ be the Laplace transforms of the notch shape function f and the flow rate function r, respectively. Show that

$$8\sqrt{\pi}s^{-1/2}F(s) = sR(s).$$

Conclude that

$$8\int_0^h \frac{f(y)}{\sqrt{h-y}}\,dy = r'(h).$$

8. Problem Derive the result of the previous problem by using the definition of the derivative instead of the Laplace transform.

9. Problem Again, let $F(s)$ and $R(s)$ be the Laplace transforms of f and r, respectively. Show that

$$F(s) = \frac{1}{8\sqrt{\pi}}s^{-1/2}(s^2R(s)).$$

Conclude that

$$f(y) = \frac{1}{88\pi} \int_0^y \frac{r''(v)}{\sqrt{y-v}}\, dv.$$

10. Problem Derive the result of Problem 7 from that of Problem 9 without using Laplace transforms.

11. Problem Rework Problems 3 and 4 using the method suggested in Problem 9 instead of Laplace transforms.

12. Problem Show that if $f(0) = 0$, then

$$r(h) = \frac{32}{3} \int_0^h (h-y)^{3/2} f'(y)\, dy.$$

13. Computation Use the program 'flow'

(invocation : '[r, h] = flow('f', H, N, ep);')

to approximate and plot the solutions of the direct problems in Exercise 1 and Problem 2. Use various values for H, the length of the interval on which the flow rate function is to be approximated, and N the number of subintervals used in the approximation. Consider both exact shape data and shape data that is perturbed with random noise of various amplitudes. Comment on the stability of the direct problem.

14. Problem Suppose f and g are notch shapes satisfying $|f(y) - g(y)| < \epsilon$ for all $y \in [0, H]$. Let r_f and r_g be the flow rate functions corresponding to f and g, respectively. Find an upper bound for the difference in the flow rates $|r_f(h) - r_g(h)|$ for $h \in [0, H]$.

15. Computation Use the program 'notch'

(invocation : '[f, y] = notch('r', H, N, ep);')

to approximate and plot the solutions of the inverse problems in Problems 3 and 4. Construct the approximations on various intervals, using various numbers of subintervals for the approximations. Consider both exact flow data and flow data perturbed with random noise. Comment on the instability of the inverse problem.

4.6.3 Notes and Further Reading

The weir notch problem is treated in W. C. Brenke's article "An application of Abel's integral equation," *American Mathematical Monthly* **29** (1922), pp. 58–

60, and in C. W. Groetsch, "Inverse problems and Torricelli's law," *The College Mathematics Journal* **24** (1993), pp. 210–217. An analysis of weir notches from the hydraulic engineer's viewpoint can be found in J. K. Vennard's book, *Elementary Fluid Mechanics* (Wiley, 1940).

5

Inverse Problems in Linear Algebra

The basic problem of linear algebra—solving a system of linear equations—is an inverse problem. In courses in linear algebra, considerable attention is given to the issues of existence and uniqueness of solutions, but the third issue raised in inverse problems, instability, is typically given short shrift. In the context of linear algebra, this instability of inverse problems manifests itself in the form of ill-conditioning of linear systems. The problem of solving a linear system is the type of inverse problem that we called a causation problem in Chapter 1. The other type of inverse problem, the identification problem, involves fundamental ideas from linear algebra (linear independence and matrix representation), yet these fundamental notions are seldom presented in the context of inverse problems.

The first module of this chapter reconsiders a number of fundamental concepts from linear algebra within the framework of inverse problems. The remaining modules deal with various inverse problems in which linear algebra plays a dominant role. These problems include elementary tomography, gravimetry, determination of dynamical parameters of binary dynamical systems, stereography, and functional estimation by the Backus–Gilbert method. The mathematical topics used include projections onto hyperplanes, multiple integration, Laplace transforms, matrix inverses and generalized inverses, least-squares solutions, and eigenpair analysis. Several MATLAB programs are provided to allow the student to carry out numerical simulations of the inverse problems with the aim not only of constructing approximate solutions but also of investigating the inherent instability of the problems.

5.1 Cause and Identity

Course Level:

Linear Algebra

Goal:

Interpret some basic problems of linear algebra as inverse problems of causation and model identification.

Mathematical Background:

Matrices and linear equations, vector and matrix norms, Gauss–Jordan elimination method

Scientific Background:

Ohm's Law, Kirchhoff's Law

Technology:

MATLAB or other high-level numerical software

5.1.1 Introduction

Linear algebra is the one course in the undergraduate curriculum in which the issues of existence, uniqueness, and stability raised by inverse problems get serious, though often inadequate, attention. The *direct* problem of linear algebra consists of determining the action of a linear transformation represented (relative to given bases) by a matrix: Given an $m \times n$ matrix A and an n-vector $\mathbf{x}$, determine the m-vector $\mathbf{b} = A\mathbf{x}$. The inverse *causation* problem, that is, the problem of finding all solutions $\mathbf{x}$ of $A\mathbf{x} = \mathbf{b}$, probably gets more attention than any other problem in elementary linear algebra. A less frequently treated inverse problem is the *identification* problem: Identify the matrix A, given an appropriate collection of "input–output" pairs $(\mathbf{x}, \mathbf{b})$ satisfying $A\mathbf{x} = \mathbf{b}$. This module is an elementary presentation of both of these inverse problems for $m \times n$ real matrices.

First we consider the inverse causation problem. A solution $\mathbf{x}$ of this problem, that is, a vector $\mathbf{x} \in R^n$ satisfying $A\mathbf{x} = \mathbf{b}$, where A is a given $m \times n$ real matrix and $\mathbf{b} \in R^m$ is a given vector, exists if and only if $\mathbf{b}$ lies in the *range* of A, that is, in the subspace

$$R(A) = \{A\mathbf{x} : \mathbf{x} \in R^n\}.$$

This subspace is, according to the definition of the action of the matrix A on a vector $\mathbf{x}$, just the subspace of R^m consisting of all linear combinations of the column vectors of A. Determining whether $\mathbf{b} \in R(A)$, that is, whether a solution exists, and finding all solutions, is accomplished by that excellent algorithm, the method of Gaussian elimination.

The uniqueness issue is addressed by a subspace of R^n associated with A, the *null-space*

$$N(A) = \{\mathbf{x} \in R^n : A\mathbf{x} = \mathbf{0}\}.$$

Again the Gaussian elimination algorithm is an effective means of characterizing the null-space and thereby settling the uniqueness question.

The stability, with respect to perturbations in the right-hand side $\mathbf{b}$ of the solution $\mathbf{x}$ of the problem $A\mathbf{x} = \mathbf{b}$, can be quantified in terms of the *condition number* of the matrix A. We assume that a unique solution exists for each $\mathbf{b}$, that is, that A is an invertible matrix. We would like to know to what extent relatively small errors in $\mathbf{b}$ can lead to relatively large changes in the solution $\mathbf{x}$. Suppose $\tilde{\mathbf{b}}$ is a perturbation of the right-hand side $\mathbf{b}$. The size of this perturbation relative to the size of $\mathbf{b}$, measured in terms of a given norm $\|\cdot\|$, is then $\|\mathbf{b} - \tilde{\mathbf{b}}\|/\|\mathbf{b}\|$. Let x be the unique solution of the system corresponding to the right-hand side $\mathbf{b}$, and let $\tilde{\mathbf{x}}$ be that corresponding to the right-hand side $\tilde{\mathbf{b}}$. Then

$$\|\mathbf{x} - \tilde{\mathbf{x}}\| = \|A^{-1}\mathbf{b} - A^{-1}\tilde{\mathbf{b}}\| \leq \|A^{-1}\|\|\mathbf{b} - \tilde{\mathbf{b}}\|;$$

hence the matrix norm $\|A^{-1}\|$ gives a bound for the change in the solution arising from a perturbation in the right-hand side. A relative measure of this change is obtained as follows:

$$\|\mathbf{x} - \tilde{\mathbf{x}}\| \leq \|A^{-1}\|\|\mathbf{b}\|\frac{\|\mathbf{b} - \tilde{\mathbf{b}}\|}{\|\mathbf{b}\|} \leq \|A^{-1}\|\|A\|\|\mathbf{x}\|\frac{\|\mathbf{b} - \tilde{\mathbf{b}}\|}{\|\mathbf{b}\|},$$

and hence

$$\frac{\|\mathbf{x} - \tilde{\mathbf{x}}\|}{\|\mathbf{x}\|} \leq cond(A)\frac{\|\mathbf{b} - \tilde{\mathbf{b}}\|}{\|\mathbf{b}\|},$$

where $cond(A) = \|A\|\|A^{-1}\|$ is called the *condition number* of the matrix A (with repect to the norm $\|\cdot\|$). The condition number therefore gives an upper bound for the relative error in the solution caused by a given relative error in the right-hand side. For matrices with large condition numbers, that is, *ill-conditioned* matrices, relatively small perturbations in the right-hand side can give rise to relatively large changes in the solution. It is in this sense that ill-conditioned systems are said to be unstable.

We now consider what can be accomplished with linear systems that have no solution, or too many solutions. In the case when $\mathbf{b} \in R^m$ is not in the range of the $m \times n$ matrix A, there is no solution to the problem $A\mathbf{x} = \mathbf{b}$, but all is not lost. A remarkable relationship between the null-space, range, and transpose allows the development of a type of generalized solution, and such generalized solutions *always* exist. The sort of generalized solution we have in mind is a *least-squares* solution, that is, a vector $\mathbf{u} \in R^n$ that minimizes the quantity $\|A\mathbf{x} - \mathbf{b}\|$ over all $\mathbf{x} \in R^n$, where the norm is the usual euclidean norm. Note that this minimum is zero if and only if the system has a solution. If $\mathbf{u}$ is a least-squares solution, then for any vector $\mathbf{v} \in R^n$, the function

$$g(t) = \|A(\mathbf{u} + t\mathbf{v}) - \mathbf{b}\|^2 = \|A\mathbf{u} - \mathbf{b}\|^2 + 2(A\mathbf{v}, A\mathbf{u} - \mathbf{b})t + \|A\mathbf{v}\|^2 t^2,$$

where $(\cdot, \cdot)$ is the familiar euclidean inner product, has a minimum at $t = 0$. The necessary condition $g'(0) = 0$ for a minimum then gives

$$(A\mathbf{v}, A\mathbf{u} - \mathbf{b}) = 0,$$

and hence $(\mathbf{v}, A^T A\mathbf{u} - A^T\mathbf{b}) = 0$ for all $\mathbf{v} \in R^n$. That is, if $\mathbf{u}$ is a least-squares solution, then

$$A^T A\mathbf{u} = A^T\mathbf{b},$$

where A^T is the transpose of A.

Conversely, if $A^T A\mathbf{u} = A^T\mathbf{b}$, then for any $\mathbf{x} \in R^n$,

$$\begin{aligned}
\|A\mathbf{x} - \mathbf{b}\|^2 &= \|A(\mathbf{x} - \mathbf{u}) + A\mathbf{u} - \mathbf{b}\|^2 \\
&= \|A(\mathbf{x} - \mathbf{u})\|^2 + 2(A(\mathbf{x} - \mathbf{u}), A\mathbf{u} - \mathbf{b}) + \|A\mathbf{u} - \mathbf{b}\|^2 \\
&= \|A(\mathbf{x} - \mathbf{u})\|^2 + 2(\mathbf{x} - \mathbf{u}, A^T A\mathbf{u} - A^T\mathbf{b}) + \|A\mathbf{u} - \mathbf{b}\|^2 \\
&\geq \|A\mathbf{u} - \mathbf{b}\|^2,
\end{aligned}$$

that is, $\mathbf{u}$ is a least-squares solution of $A\mathbf{x} = \mathbf{b}$.

So, least-squares solutions of $A\mathbf{x} = \mathbf{b}$ coincide with ordinary solutions of the symmetric problem $A^T A\mathbf{x} = A^T\mathbf{b}$. Now this symmetric problem *always* has a solution since $R(A^T) = R(A^T A)$ (see Problem 9) and hence $A^T\mathbf{b} \in R(A^T A)$ for *any* $\mathbf{b} \in R^m$. Therefore, any linear system $A\mathbf{x} = \mathbf{b}$ has a least-squares solution. However, least-squares solutions need not be unique. Indeed, if $\mathbf{u}$ is a least-squares solution, then so is $\mathbf{u} + \mathbf{v}$ for any $\mathbf{v} \in N(A)$, that is, the set of least-squares solutions forms a hyperplane parallel to the null-space. Therefore, if A has a nontrivial null-space, then $A\mathbf{x} = \mathbf{b}$ has infinitely many least-squares solutions. However, one least-squares solution can be distinguished from the others, namely, the one that is orthogonal to the null-space. There can be at most one such least-squares solution, because if $\mathbf{u}$ and $\mathbf{w}$ are both least-squares solutions that are orthogonal to $N(A)$, then $\mathbf{u} - \mathbf{w}$ is orthogonal to $N(A)$. Also, $A^T A(\mathbf{u} - \mathbf{w}) = A^T\mathbf{b} - A^T\mathbf{b} = \mathbf{0}$, and hence $\mathbf{u} - \mathbf{w} \in N(A^T A) = N(A)$ (see Problem 6). Therefore, $\mathbf{u} - \mathbf{w} \in N(A) \cap N(A)^{\perp}$, that is, $\mathbf{u} = \mathbf{w}$. On the other hand, there is always a least-squares solution that is orthogonal to the null-space (see Problem 10), and hence *any* linear system has a unique least-squares solution that is orthogonal to the null-space of the coefficient matrix. If we agree to accept this notion of generalized solution, then *every* linear system has a unique (generalized) solution.

Finally, we briefly consider the identification problem, that is, the inverse problem of determining an $m \times n$ matrix A, given pairs of vectors (x, b) related by $A\mathbf{x} = \mathbf{b}$. For each such pair we call $\mathbf{x}$ the input and $\mathbf{b}$ the corresponding output. Our job is to identify the "black box" A, by "interrogating" it with appropriate inputs $\mathbf{x}$ and observing the outputs $\mathbf{b}$. Because we control the inputs, we can arrange it so that they are linearly independent, and we will assume that this has been done. It is convenient to express things in matrix form by aggregating the independent inputs $\mathbf{X}_1, \mathbf{X}_2, \ldots, \mathbf{X}_p$ as the column vectors of an $n \times p$ matrix $\mathbf{x}$, and similarly thinking of the corresponding outputs $\mathbf{B}_1, \mathbf{B}_2, \ldots, \mathbf{B}_p$ as the column vectors of an $m \times p$ matrix B. We say that A is *identifiable* from the matrix pair (X, B) if there is a unique $m \times n$ matrix A satisfying $AX = B$.

We consider three cases, each premised on different relative sizes for n and p. First note that $p > n$ is impossible, since $\{\mathbf{X}_1, \ldots, \mathbf{X}_p\}$ is a linearly independent set in R^n. If $p = n$, then X is invertible and A is identifiable, in fact, $A = BX^{-1}$. In this case a simple modification of the Gauss–Jordan elimination method provides an algorithm for identifying A. The method rests on the observation that if $AX = B$, then $X^T A^T = B^T$. One can therefore solve for the transpose of the model matrix A by augmenting the transpose of the input matrix with the transpose of the output matrix and performing the Gauss–Jordan algorithm in the usual way:

$$\left[X^T : B^T\right] \to \cdots \to \left[I : A^T\right].$$

In the last case, when $p < n$, one suspects that the input–output information is insufficient to identify A. This is indeed so since in this case there is a vector $\mathbf{q} \in R^n$ that is orthogonal to $\mathbf{X}_1, \ldots, \mathbf{X}_p$. Let C be the $m \times n$ matrix whose first row is $\mathbf{q}^T$ and whose other rows are zero vectors. Then CX is the $m \times p$ zero matrix. Therefore, if $AX = B$, then, likewise, $(A + C)X = B$, and hence A is not identifiable from the information (X, B).

5.1.2 Activities

1. Problem A company manufactures three types of circuit boards. Each board consists of three types of components, say, diodes, transistors, and resistors. Board 1 requires two diodes, seven transistors, and three resistors. Board 2 requires three diodes, five transistors, and two resistors. Board 3 requires one diode, nine transistors, and four resistors. Each of the components has a certain unit cost. Is it possible for the costs (in some monetary unit) of the constituents of the boards to be 24, 20, and 15 (for diodes, transistors, and resistors, respectively)?

2. Problem Referring to Problem 1, suppose the constituent costs of the three types of boards are 24, 20, and 28 dollars, respectively, and that the sum of the unit costs of the three types of components is a minimum. Estimate (to the nearest cent) the unit costs of the components.

3. Question Does small-determinant imply ill-conditioned?

4. Problem Given an arbitrarily small positive number ϵ, construct a matrix A with $\det A = \epsilon$ and $cond(A) = 1 + \epsilon$.

5. Computation Suppose the finite Laplace transform

$$F(s) = \int_0^1 e^{-su} f(u)du, \quad 0 \leq s \leq 1$$

is discretized to produce the $n \times n$ matrix A with $a_{ij} = e^{-ij/n^2}$, for $i, j = 1, \ldots, n$. Find $cond(A)$ for $n = 10, 20, 50$. For each such N, generate an n-vector with random components in $[-1, 1]$ and compute the vector $\mathbf{b} = A\mathbf{x}$. Plot the vector $\mathbf{x}$ and the vector $\mathbf{b}$. Explain the results.

6. Problem Show that $N(A) = N(A^T A)$ for any real matrix A.

7. Problem Show that for any real matrix A, $R(A^T)^{\perp} = N(A)$.

8. Problem Show that if W and V are subspaces of R^n and $W^{\perp} = V^{\perp}$, then $W = V$.

9. Problem Use Problem 8 to show that $R(A^T) = R(A^T A)$ for any real matrix A.

10. Problem Suppose $\mathbf{u}$ is a least-squares solution of $A\mathbf{x} = \mathbf{b}$, and let $P\mathbf{u}$ be the orthogonal projection of $\mathbf{u}$ onto $N(A)$ (i.e., $P\mathbf{u} = \sum_{j=1}^{k}(\mathbf{u}, \mathbf{v}^{(j)})\mathbf{v}^{(j)}$, where $\{\mathbf{v}^{(1)}, \ldots, \mathbf{v}^{(k)}\}$ is an orthonormal basis for $N(A)$). Show that $\mathbf{u} - P\mathbf{u}$ is a least-squares solution that is orthogonal to $N(A)$.

11. Exercise Suppose $\mathbf{b} = [1, 0, 2]^T$ and

$$A = \begin{bmatrix} 1 & 1 \\ 2 & 0 \\ 1 & 1 \end{bmatrix}.$$

Show that the system $A\mathbf{x} = \mathbf{b}$ has no ordinary solution, but that it does have a unique least-squares solution.

12. Exercise Suppose $\mathbf{b} = [1, 0, 1]^T$ and

$$A = \begin{bmatrix} 1 & 1 \\ 1 & 1 \\ 1 & 1 \end{bmatrix}.$$

Show that the system $A\mathbf{x} = \mathbf{b}$ has no ordinary solution, but that it does have infinitely many least-squares solutions. Find the least-squares solution that is orthogonal to the null-space.

13. Exercise Suppose B is a real, symmetric matrix. A given nonzero $n \times 1$ vector $\mathbf{y}$ is an eigenvector of B if the n equations in the single unknown μ

$$\mathbf{y}\mu = B\mathbf{y}$$

have an ordinary solution. Show that for each nonzero $\mathbf{y}$ this equation has the unique least-squares solution

$$\mu = \frac{(B\mathbf{y}, \mathbf{y})}{\|\mathbf{y}\|^2}.$$

This quantity is called the *Rayleigh quotient* for $\mathbf{y}$, and it may be taken as an approximation to an eigenvalue of B.

14. Problem Show that there are infinitely many 2×3 matrices that produce the outputs $[1, 1]^T$ and $[0, 1]^T$, given the respective inputs $[1, 1, -1]^T$ and $[-1, 1, 0]^T$.

15. Problem Give an example of 2×2 matrices X and B for which no matrix A exists satisfying $AX = B$.

16. Exercise Identify the matrix A satisfying $AX = B$, where

$$X = \begin{bmatrix} 1 & 1 & 1 \\ -1 & 1 & 0 \\ 0 & 1 & 1 \end{bmatrix}, \quad B = \begin{bmatrix} -2 & 6 & 3 \\ -1 & 0 & 0 \end{bmatrix}.$$

17. Problem A hydrocarbon is a compound consisting of hydrogen and carbon. Suppose $[b_1, b_2]^T$ represents the number of hydrogen and carbon atoms, respectively, in a mixture consisting of a certain combination of molecules of three hydrocarbons. Let $[x_1, x_2, x_3]^T$ represent the number of molecules, respectively, of three unknown hydrocarbons. Show that if the column vectors of the 3×3 matrix A represent, in an appropriate sense, the chemical formulas of the three hydrocarbons, then $A\mathbf{x} = \mathbf{b}$. Identify (by chemical name) the three hydrocarbons, given the input–output information (X, B), where:

$$X = \begin{bmatrix} 100 & 0 & 100 \\ 50 & 100 & 0 \\ 100 & 100 & 100 \end{bmatrix}, \quad B = \begin{bmatrix} 1500 & 1000 & 1400 \\ 1700 & 1200 & 1600 \end{bmatrix}.$$

18. Problem A refinery uses two types of crude oil, call them type 1 and type 2, to produce two products, say heating oil and gasoline. Assume that all of the ingredients (the crude oils) are used in the process and none is wasted. Furthermore, assume that the products (the number of barrels of heating oil and gasoline) are linear functions of the ingredients (the number of barrels of each type of crude oil). Suppose it is observed that when 4,000 barrels of type 1 crude and 1,000 barrels of type 2 crude are used, 1,900 barrels of heating oil and 3,400 barrels of gasoline are produced. However, if 2,000 barrels of type 1 crude and 3,000 barrels of type 2 crude are used, 2,700 barrels of heating oil and 2,200 barrels of gasoline are produced. What are the "recipes" for the two types of product?

19. Problem Show that if $\{\mathbf{X}_1, \ldots \mathbf{X}_n\} \subseteq R^n$ are orthonormal inputs with corresponding outputs $\{\mathbf{B}_1, \ldots, \mathbf{B}_n\} \subseteq R^m$, then (X, B) identifies the model $A = BX^T$.

20. Problem Suppose $AX = B$, where A is an $m \times n$ matrix and X is an $n \times p$ matrix, with $p > n$. Show that if $rank(XX^T) = n$, then A is identifiable from (X, B). What is A in terms of X and B?

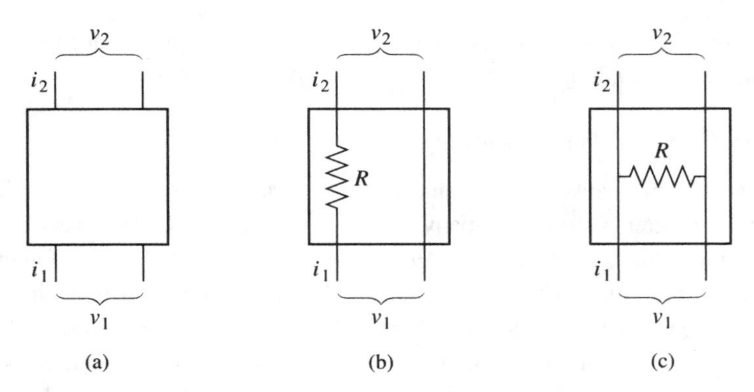

Figure 5.1: Series and Shunt Resistances

21. Problem This problem is concerned with the identification of "black boxes" representing certain very simple electrical circuits. The input to the box (indicated on the underside of part (a) of Figure 5.1) is a vector $[v_1, i_1]^T$ representing an input voltage and current, respectively, and the output is a similar vector $[v_2, i_2]^T$.

(a) Consider first the case of a single resistor R in series, pictured in part (b) of Figure 5.1. Use Ohm's Law (voltage drop = resistance $\times$ current) to show that the model matrix for the series circuit is

$$\begin{bmatrix} 1 & -R \\ 0 & 1 \end{bmatrix}.$$

(b) A shunt resistance R is pictured in part (c) of Figure 5.1. Show that the model matrix for a shunt is

$$\begin{bmatrix} 1 & 0 \\ -\frac{1}{R} & 1 \end{bmatrix}.$$

(For this you will need Kirchhoff's Law: The directed sum of currents at a node is zero.)

(c) We will call a series resistance followed by a shunt resistance a Γ-circuit, and a shunt resistance followed by a series resistance will be called an L-circuit (draw the pictures to appreciate the notation). Find the model matrices for a Γ-circuit and for an L-circuit.

(d) Show that observation of a single input–output pair is sufficient to identify a Γ-circuit (L-circuit) if the circuit is known to be a Γ-circuit (L-circuit), but that if a circuit is known to be either a Γ-circuit or an L-circuit, without knowing which is the case, two linearly independent inputs, and the corresponding outputs, are necessary to identify the circuit.

5.1.3 Notes and Further Reading

Background material on elementary linear algebra appropriate for the topics in this module can be found in many sources, for example C. W. Groetsch and J. T. King, *Matrix Methods and Applications*, Prentice-Hall, 1988. Given an $m \times n$ matrix A, there is an $n \times m$ matrix $A^\dagger$ that associates with each vector $\mathbf{b} \in R^m$ the unique least-squares solution of $A\mathbf{x} = \mathbf{b}$ that is orthogonal to $N(A)$. The matrix $A^\dagger$ is called the Moore–Penrose pseudoinverse of A and it can be computed in MATLAB simply by issuing the command "pinv(A)." More on the Moore–Penrose pseudoinverse can be found in, for example, G. Strang, *Introduction to Linear Algebra*, Wellesley-Cambridge Press, 1993.

5.2 L'ART Pour L'Art

Course Level:

Linear Algebra

Goal:

Develop a geometrically based iterative method for elementary tomography.

Mathematical Background:

Projections, hyperplanes, linear equations

Scientific Background:

None

Technology:

MATLAB

5.2.1 Introduction

The historical introduction to these notes began with Plato's problem of reconstructing reality from shadows projected on a wall and ended with Cormack's development of computed tomography. Both of these problems demand the reconstruction of an object from knowledge of certain projections, which may be regarded as (generalized) averages through slices of the object. Generally such problems are underdetermined and unique reconstruction of the object is impossible, given the meager set of data. Nevertheless, methods for constructing approximate solutions that are consistent with the data can be very useful in a number of technical settings. In this module we discuss a simple geometrically motivated algorithm, the *algebraic reconstruction technique* (ART), for approximating a solution of a very basic linear algebra problem arising in tomography.

The ART algorithm uses successive orthogonal projections to approximate solutions of systems of linear equations. To see how ART works, consider the simplest example of a uniquely solvable system of two linear equations with two unknowns. Such a system is represented geometrically by two intersecting lines in the plane. Take any point in the plane and project it onto the first line (that is, drop a perpendicular from the point to the first line). Now project the resulting point onto the second line, then project that point onto the first line, and so on. Continue projecting alternately onto the two lines and watch as the path zig-zags into the point of intersection of the two lines.

Now take the case of a single linear equation in k unknowns:

$$v_1 x_1 + v_2 x_2 + \cdots + v_k x_k = \mu.$$

For $k = 2$ the solution set of this equation is a line, for $k = 3$ it is a plane, and in general the solution set is called a hyperplane. This hyperplane is determined by the coefficient vector $\mathbf{v}$ and the scalar μ, and it can be expressed as

$$H = \{\mathbf{x} \in R^k : (\mathbf{v}, \mathbf{x}) = \mu\},$$

where $(\cdot, \cdot)$ is the usual inner product. The vector $\mathbf{v}$ is normal to the hyperplane since $(\mathbf{v}, \mathbf{x} - \mathbf{z}) = 0$ for all $\mathbf{x}, \mathbf{z} \in H$. For a given $\mathbf{y} \in R^k$, the orthogonal projection of $\mathbf{y}$ onto H is therefore the unique vector $P\mathbf{y} \in H$ with $\mathbf{y} - P\mathbf{y} = t\mathbf{v}$, for some scalar t. Since $P\mathbf{y} \in H$, the scalar t must satisfy

$$t\|\mathbf{v}\|^2 = (\mathbf{v}, \mathbf{y} - P\mathbf{y}) = (\mathbf{v}, \mathbf{y}) - \mu.$$

But $P\mathbf{y} = \mathbf{y} - t\mathbf{v}$, and hence

$$P\mathbf{y} = \mathbf{y} + \frac{\mu - (\mathbf{v}, \mathbf{y})}{\|\mathbf{v}\|^2}\mathbf{v}.$$

Therefore, projecting onto a hyperplane is a very easy operation to perform, particularly if the vector $\mathbf{v}$ has relatively few nonzero components.

Any solution of a system of m linear equations in k unknowns lies in the intersection of the m hyperplanes determined by the individual equations. The ART algorithm approximates a solution by projecting a given vector successively onto these m hyperplanes. A single cycle of the method applied to a vector $\mathbf{y}$ gives a new approximation

$$\mathbf{y}^{(1)} = Q\mathbf{y}, \quad Q = P_m P_{m-1} \dots P_1,$$

where P_j is the projector onto the jth hyperplane. Repeated cycles then give further approximations, $\mathbf{y}^{(2)} = Q\mathbf{y}^{(1)}$, etc.

We now apply ART to a very simple tomography problem. Consider an object to be an $n \times n$ array of "pixels." We assume that the object is homogeneous within any given pixel, but its composition might change from pixel to pixel. Our goal is to get a picture of this variability from pixel to pixel. Suppose the object is illuminated with a beam of radiation that passes through the centers of some of the pixels. We order the pixels from left to right starting with the top row, as illustrated in the 3×3 example in Figure 5.2. The path of a beam through the object is specified by a *view vector*, that is, a row vector of dimension n^2 whose components are 1 or 0 depending on whether the beam intersects the pixel or not. For example, in Figure 5.2 the view vector of the beam pictured is

$$\mathbf{v} = [0, 1, 0, 1, 0, 0, 0, 0, 0].$$

Suppose the material within a pixel absorbs a fraction of the radiation that is incident upon it. This fraction is a number in the interval $(0, 1)$, and a value close to 0 corresponds to a nearly transparent pixel, while a value close to 1 indicates a nearly opaque pixel. We will call the fraction p_j absorbed by the jth pixel the *absorption coefficient* of that pixel. Suppose a beam with view vector

$$\mathbf{v} = [v_1, v_2, \dots, v_k], \quad v_j \in \{0, 1\}, \quad j = 1, \dots, k = n^2$$

passes through an object with absorption coefficients

$$[p_1, p_2, \dots, p_k].$$

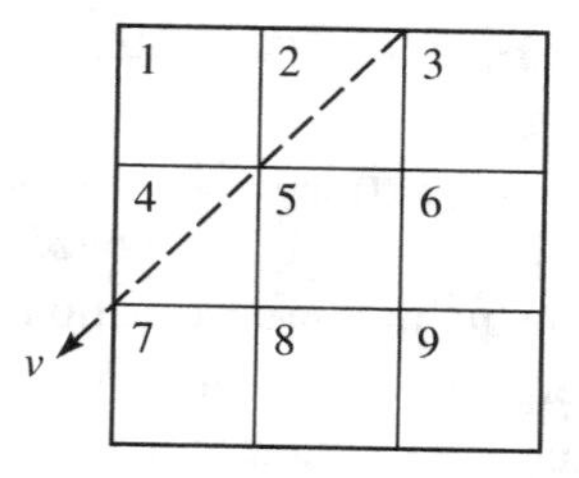

Figure 5.2: A View of an Object

If the beam passes through the jth pixel, then a fraction p_j of the radiation is absorbed and a fraction $1 - p_j$ is passed on. The fraction p of the radiation that emerges from the object is then

$$p = (1 - p_1)^{v_1}(1 - p_2)^{v_2} \dots (1 - p_k)^{v_k},$$

where $\mathbf{v}$ is the view vector of the beam, or, equivalently,

$$v_1 \ln(1 - p_1) + v_2 \ln(1 - p_2) + \cdots + v_k \ln(1 - p_k) = \ln p.$$

Note that each of the numbers $\ln(1 - p_j)$, $\ln p$, is negative, so we prefer to write this relationship as

$$v_1 x_1 + v_2 x_2 + \cdots + v_k x_k = \mu,$$

where the numbers

$$x_j = -\ln(1 - p_j), \quad \mu = -\ln p = \sum\{x_j : v_j = 1\}$$

are positive. The tomography problem is therefore equivalent to finding a point in the intersection of the m hyperplanes determined by the m view vectors and corresponding right-hand sides. Furthermore, for a given view the right-hand side is the sum of those components x_j associated with pixels through which the beam passes.

Given m view vectors $\mathbf{v}^{(1)}, \dots, \mathbf{v}^{(m)}$ and corresponding values $\mu_1, \dots, \mu_m$, we collect the views as row vectors of a *view matrix* V, whose ijth entry is $v_j^{(i)}$, and collect the measurements μ_j corresponding to the respective views into an m-dimensional row vector mu. Given an n^2-dimensional row vector $\mathbf{x}^{(0)}$ representing an initial approximation, any number N of complete projection cycles of the ART algorithm can be carried out to approximate a solution vector $\mathbf{x}$. The program 'art1' (invocation: [x]=art1(V,mu,x0,N)) accomplishes this with the extra twist that after each cycle the approximate solution is projected onto the positive orthant (recall that $x_j \geq 0$ for all j). The approximate solution provided by 'art1' can be displayed as an $n \times n$ grid with appropriately shaded pixels by the program 'displa'.

Here is a simple example of the tomographic coding technique. Consider a 3×3 object representing an "X" with a nearly black main diagonal and a gray alternate diagonal. We might code this as in Figure 5.3, with the vector representing the object being

$$\mathbf{x} = [1, 0, .5, 0, 1, 0, .5, 0, 1].$$

1	0	.5
0	1	0
.5	0	1

Figure 5.3: A Sample Object

Suppose we take six views of the object consisting of slices across the three rows, down the main diagonal, up the other diagonal, and down the middle column. The view matrix for these observations is then

$$V = \begin{bmatrix} 1 & 1 & 1 & 0 & 0 & 0 & 0 & 0 & 0 \\ 0 & 0 & 0 & 1 & 1 & 1 & 0 & 0 & 0 \\ 0 & 0 & 0 & 0 & 0 & 0 & 1 & 1 & 1 \\ 1 & 0 & 0 & 0 & 1 & 0 & 0 & 0 & 1 \\ 0 & 0 & 1 & 0 & 1 & 0 & 1 & 0 & 0 \\ 0 & 1 & 0 & 0 & 1 & 0 & 0 & 1 & 0 \end{bmatrix},$$

and the corresponding measurement vector is

$$\text{mu} = [1.5, 1, 1.5, 3, 2, 1].$$

5.2.2 Activities

1. Exercise Two students take a reading class that requires that each student write two papers. The professor has lost the grade sheet for the course, but she remembers the class average score on each paper and the average scores over the two papers for each student. Can she reconstruct the individual scores on each paper?

2. Problem Suppose unknown weights are associated with the three vertices of a regular triangle. For each vertex the sum of the weights at the adjacent vertices is known. Show that the weight at each vertex is uniquely determined. Does a similar result hold for a quadrilateral? Pentagon? N-gon?

3. Exercise Suppose unknown weights are associated with the vertices of a cube. Are the weights uniquely determined from knowledge, at each vertex, of the sum of the weights at the adjacent vertices?

4. Problem Given $\mathbf{y} \in R^k$, show that the vector

$$\mathbf{q} = \mathbf{y} + \frac{\mu - (\mathbf{v}, \mathbf{y})}{||\mathbf{v}||^2}\mathbf{v}$$

is the orthogonal projection of $\mathbf{y}$ onto the hyperplane $H = \{\mathbf{x} : (\mathbf{x}, \mathbf{v}) = \mu\}$ by showing that (a) $\mathbf{q} \in H$ and (b) $\mathbf{q}$ is the nearest point in H to $\mathbf{y}$.

5. Exercise Perform two cycles of the ART algorithm, using initial approximation $[1, 0]$ and working to three significant digits, on the linear system

$$-x_1 + 2x_2 = 1$$

$$3x_1 + x_2 = 2.$$

6. Problem Show that if $ac + bd = 0$, $ab \neq 0$, $cd \neq 0$, then at most two projections in the ART algorithm are sufficient to solve the system

$$ax_1 + bx_2 = \mu_1$$

$$cx_1 + dx_2 = \mu_2.$$

7. Problem Find two distinct 3×3 objects that give the same measurement vector for the six views corresponding to the row sums, the main diagonal sum, the other diagonal sum, and the middle column sum.

8. Exercise For the views indicated in Figure 5.4 for the 3×3 object, construct the view matrix V, the measurement vector mu, and the solution vector $\mathbf{x}$. (Use weight 1 for dark pixels and 0 for light pixels.)

9. Computation A worm has burrowed into an apple modeled as a 5×5 grid. Horizontal probes through the rows (top to bottom) yield measurements of 0, 2, 1, 2, 0, respectively, while vertical probes down the columns (left to right) yield values of 0, 1, 3, 1, 0, respectively. The main diagonal (NW to SE)

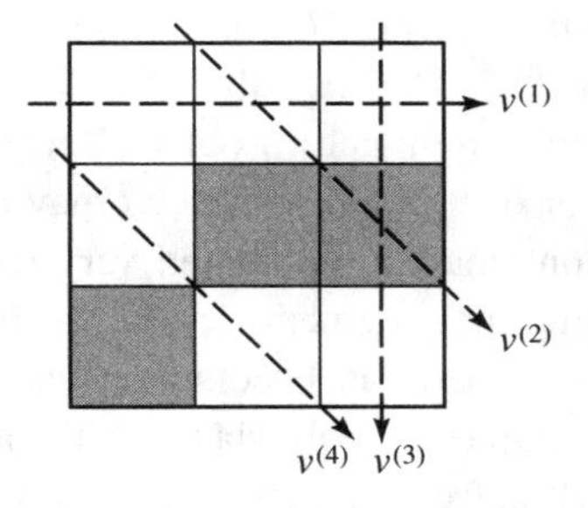

Figure 5.4: Object for Exercise 8

sum is 1, while the other diagonal (SW to NE) sum is 3. Use 'art1' and 'displa', with the zero vector as an initial approximation, to get a picture of the worm.

10. Computation A 6×6 object has row sums (top to bottom) 0, 2, 0, 2, 6, 0 and column sums (left to right) 2, 2, 1, 1, 2, 2. Try to reconstruct the object using 'art1' and 'displa' using the zero vector as an initial approximation. Now add four more views and measurements consisting of a ray through pixels 4, 11, and 18 with sum 1; a ray through pixels 24, 29, and 34 with sum 2; a ray through pixels 19, 26, and 33 with sum 2; and a ray through pixels 3, 8, and 13 with sum 1. Try to reconstruct the picture using this additional information, and compare the result with the previous reconstruction.

11. Computation Repeat the previous computations using various (nonzero) initial approximations, and compare the results with the previous reconstructions.

12. Computation Repeat the previous computation, blending 5 percent uniform random noise into the measurements, and compare the results to the previous reconstructions.

5.2.3 Notes and Further Reading

The word "tomography" is based on the Greek root "tomos" meaning a cut or slice. What we have called "views" in this module could be called slices through the object. The shadowy images sometimes seen in the Activities above arise from the underdetermined nature of the linear systems involved. In the tomography community such spurious images are called "ghosts."

The ART algorithm is not new; it dates back to the work of Stefan Kaczmarz in the mid-1930s (Kaczmarz was murdered in a Nazi roundup of intellectuals following the invasion of Poland in 1939). A convergence proof of ART, in a more general context than that of this module, can be found, for example, in C. W. Groetsch, *Inverse Problems in the Mathematical Sciences*, Vieweg, Braunschweig, 1993. The ART algorithm has a number of advantages over direct methods for the tomography problem. Since all components of view vectors are either zero or one, the view vectors may be stored as bit strings, and individual projections may be computed very quickly. Also, new view vectors and corresponding measurements can be easily introduced during the course of the computation if new data becomes available. Furthermore, a priori information can be incorporated simply via the initial approximation vector.

"Image Reconstruction from Projections," by R. Gordon, G. Herman, and S. Johnson, *Scientific American*, Vol. 233 (October 1975), pp. 56–68, is an excellent popular article on computed tomography that gives more details on the medical technology involved. Ivars Peterson's article, "Inside Averages," *Science News* (May 1986), pp. 300–301, discusses an interesting application of tomography to literature.

5.3 Nonpolitical Pull

Course Level:

Linear Algebra

Goal:

Investigate the instability of a model problem in geophysics.

Mathematical Background:

Midpoint rule, matrix inverses, eigenvalues

Scientific Background:

Inverse-square law of gravity

Technology:

Graphics–symbolic calculator, MATLAB

5.3.1 Introduction

Let's renew our acquaintance with the Rhine maidens (see the module *das Rheingold*). But now, instead of a discrete nugget of gold, we wish to identify a nonhomogeneous mass density $w(s)$, $0 \le s \le 1$. Such a mass distribution engenders an inhomogeneity $\mu(x)$ in the vertical force of gravity at the surface. The relationship between w and μ is easily obtained as in the earlier module and is illustrated in Figure 5.5: The vertical component of force $\Delta\mu(x)$ at position

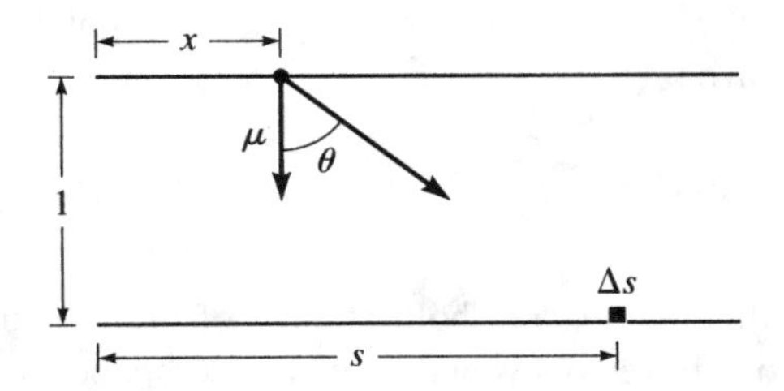

Figure 5.5: Gravitational Attraction of a Distributed Mass

x engendered by a mass segment of length Δs at position s on the subsurface is

$$\begin{aligned}\Delta\mu(x) &= \gamma w(s)((x-s)^2+1)^{-1}\cos\theta\Delta s\\ &= \gamma((x-s)^2+1)^{-3/2}w(s)\Delta s,\end{aligned}$$

where $w(s)$ is the mass density at position s. The usual summation and limit process leads to the model

$$\mu(x) = \int_0^1 ((x-s)^2+1)^{-3/2}w(s)ds$$

(here, and henceforth, we take $\gamma = 1$ for convenience). The problem of determining the gravitational inhomogeneity μ from the mass density w is a straightforward *direct* problem. On the other hand, the problem of determining the inaccessible mass density w from a known (i.e., measured) gravitational inhomogeneity μ is a classic *inverse* problem.

A notable feature of the model above is that μ is generally smoother than w. Even for quite "rough" (e.g., discontinuous) functions w, the function μ is infinitely differentiable because it inherits its smoothness in x from the kernel function $((x-s)^2+1)^{-3/2}$. In "filtering" the function w through the integral we can expect some of the fine detail in w to be "smoothed out" in the process. The essential point is that μ contains less information than w and we can therefore expect that the inverse problem of reconstructing w from knowledge of μ will be difficult.

We can form a concrete appreciation for the difficulties involved by studying an approximating discrete problem. One way of doing this comes about if we replace the integral by an approximate quadrature rule. For example, we might use the *midpoint rule*, that is,

$$\int_0^1 f(s)\,ds \approx h\sum_{j=1}^{n} f(s_j),$$

where $h = 1/n$ and $s_j = (j-\frac{1}{2})h$ for $j = 1,\dots,n$. Applying this rule to the model above we have

$$\mu(x) \approx h\sum_{j=1}^{n}((x-s_j)^2+1)^{-3/2}w(s_j).$$

If we insist that this relationship hold at each of the midpoints $x = s_i$ for $i = 1,\dots,n$, we obtain vectors $\mathbf{w}$ and $\boldsymbol{\mu}$, whose components approximate the values of the functions w and μ, respectively, at the midpoints of the subintervals formed in the discretization process. The vectors $\mathbf{w}$ and $\boldsymbol{\mu}$ are related by the matrix equation

$$A\mathbf{w} = \boldsymbol{\mu},$$

where A is the $n \times n$ matrix with entries

$$a_{ij} = h((s_i - s_j)^2 + 1)^{-3/2} \quad i, j = 1, \ldots, n.$$

The *discrete* model

$$A\mathbf{w} = \boldsymbol{\mu}$$

may then be taken as an approximation to the *continuous* model

$$\int_0^1 ((x - s)^2 + 1)^{-3/2} w(s)\, ds = \mu(x).$$

The program 'geo' will produce, for a given positive integer n, the $n \times n$ matrix A that is the discrete model of the geophysical prospecting problem, along with the vector $\mathbf{s}$ of midpoint samples. It is then easy and entertaining to visualize the dramatic smoothing properties of the discrete model. For example, in Figure 5.6 a very rough mass density on 50 midpoints is plotted that consists of random noise in the interval $[0, 2]$. Applying the matrix A, obtained from

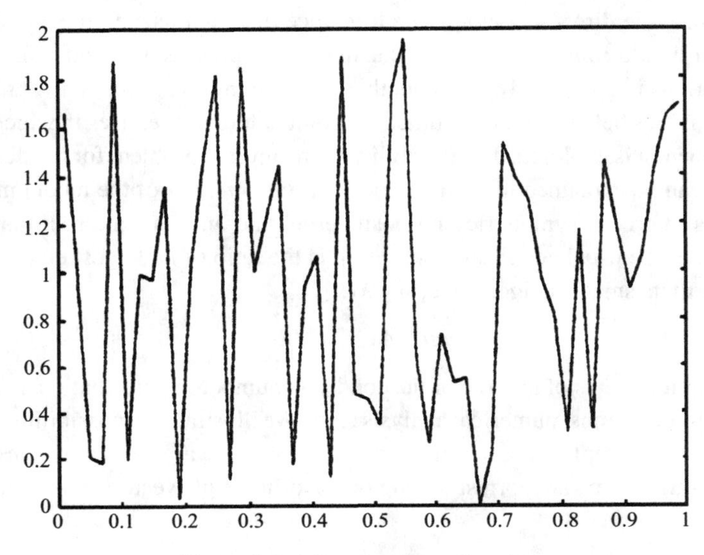

Figure 5.6: A Random Mass Density

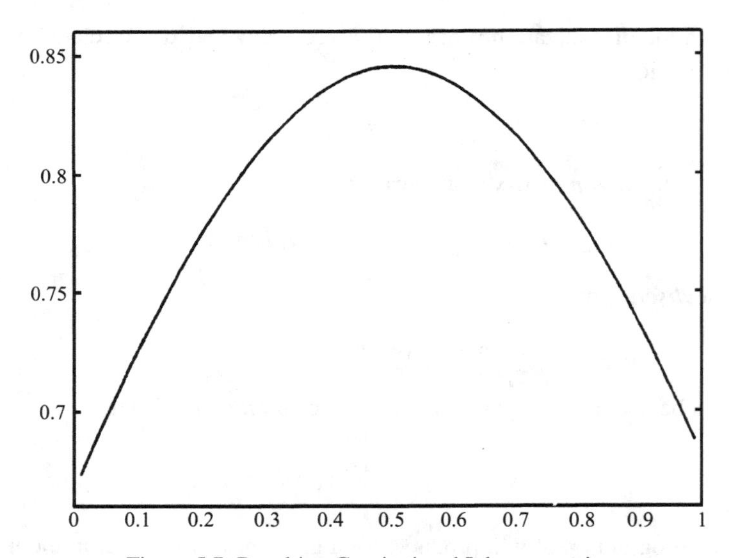

Figure 5.7: Resulting Gravitational Inhomogeneity

'geo' with $n = 50$, to this discrete density results in the very smooth curve plotted in Figure 5.7. The plot in the second figure represents the smooth, information-poor function μ, while that in the first figure illustrates the rough, information rich density w that engenders μ.

Since the direct model appears to reduce fluctuations, i.e., to smooth the input, it should come as no surprise that the inverse process will tend to magnify fluctuations in the data. To put it another way, the inverse problem is *unstable*. The activities below contain a number of computational exercises in which this phenomenon is explored. The instability of the inverse problem for the discrete model can be explained in terms of the *condition number* of the model matrix A. This matrix is symmetric and positive definite and its condition number, relative to the usual euclidean norm $\|\cdot\|$, is the ratio of its largest eigenvalue, say λ_n, to its smallest eigenvalue, say λ_1:

$$cond(A) = \lambda_n/\lambda_1.$$

General treatments of the role of the condition number in perturbation analysis can be found in most numerical analysis texts. We illustrate the possibilities with a particular example. Suppose that $0 < \lambda_1 < \cdots < \lambda_n$ are the eigenvalues of A and that $\mathbf{u}_1, \ldots, \mathbf{u}_n$ are corresponding orthonormal eigenvectors. If $\boldsymbol{\mu} = \lambda_n \mathbf{u}_n$, then

$$A\mathbf{w} = \boldsymbol{\mu},$$

where $\mathbf{w} = \mathbf{u}_n$. If the right-hand side of this equation is perturbed by

$$\Delta \boldsymbol{\mu} = \frac{a}{\lambda_1}\mathbf{u}_1$$

for some positive scalar a, then the solution of the perturbed system is

$$\mathbf{w} + \Delta \mathbf{w} = \mathbf{u}_n + \frac{a}{\lambda_1}\mathbf{u}_1.$$

The relative sizes of the perturbations are then related in the following way:

$$\begin{aligned}\frac{\| \Delta \mathbf{w} \|}{\| \mathbf{w} \|} &= \frac{\lambda_n}{\lambda_1}\frac{\| \frac{a}{\lambda_1}\mathbf{u}_1 \|}{\| \lambda_n \mathbf{u}_n \|} \\ &= cond(A)\frac{\| \Delta \boldsymbol{\mu} \|}{\| \boldsymbol{\mu} \|}.\end{aligned}$$

Therefore, the relative error in the right-hand side may be magnified by a factor of $cond(A)$. Linear systems with coefficient matrices having large condition numbers are called *ill-conditioned*. Unstable inverse problems give rise to discrete models with ill-conditioned matrices. The solution of such ill-conditioned systems is particularly challenging as the data of the problem, which is represented in the right-hand side of the discrete model, invariably contains errors. Even if the error is only that which results from representing the real numbers in a computer in floating-point form, a severely ill-conditioned matrix can have very unpleasant effects, as will be seen in some of the computations below. In general, data errors are greatly magnified in the solution process for ill-conditioned problems, and special measures must be taken to dampen this error magnification. One method for accomplishing this is hinted at in the activities.

5.3.2 Activities

1. Computation Generate the discrete models A for $n = 5, 10, 20$, and 40, and in each case use the MATLAB function 'cond' to find the condition number of A.

2. Computation Use the program 'geo' to produce the 100×100 discrete model matrix A. Generate various random 100-vectors $\mathbf{x}$, plot $\mathbf{x}$ and $A\mathbf{x}$, and note the qualitative features of both.

3. Computation Show that the matrix A of the discrete model has positive eigenvalues for $n = 5, 10,$ and 20.

4. Problem Find the gravitational inhomogeneity μ engendered by the constant mass density $w(s) = 1$ (this is the density used in the program 'geo').

5. Computation Use 'geo' to discretize the integral operator forming the 50×50 discrete model A. Use MATLAB to solve the discrete problem and plot the resulting computed mass density. Plot the true mass density (obtained in the previous problem) and compare.

6. Problem Suppose that $\alpha = k\lambda_1$ where λ_1 is the smallest eigenvalue of the positive definite matrix A. Show that

$$cond(A + \alpha I) \leq \frac{cond(A)}{k+1} + 1,$$

where I is the identity matrix. Does this suggest a way of decreasing the ill-conditioning of the discrete model at the expense of increasing the modeling error?

7. Computation Repeat the computation in Activity 5, but now perturb the matrix A with a small positive multiple of the identity matrix before solving the discrete system. Experiment with various positive multiples, plotting the results each time. Repeat the experiments by polluting the data with random error through the use of the parameter 'ep' in the program 'geo'.

8. Calculation Calculate the gravitational inhomogeneity engendered by the mass distribution $w(s) = s(1 - s)$.

9. Project Modify the program 'geo' to accommodate the functions from the previous activity. Repeat Computation 7 for this new model.

10. Calculation Find the gravitational inhomogeneity engendered by the density $w(s) = s^2(1 - s)^2$.

11. Problem Consider a mass that is distributed on a plane that is parallel to the horizontal surface plane and z units below the surface plane. Let $w(s, t)$ be the mass density at a point (s, t) on the subsurface plane. Show that the vertical gravitational inhomogeneity at a point (x, y) on the surface is

$$\mu(x, y) = z\gamma \int_{-\infty}^{\infty} \int_{-\infty}^{\infty} ((x - s)^2 + (y - t)^2 + z^2)^{-3/2} w(s, t)\, ds\, dt.$$

12. Problem Let $h(x, y, r)$ be the average value of the mass density in the previous problem on a circle of radius r in the subsurface plane centered on the point $(x, y, -z)$, that is

$$h(x, y, r) = \frac{1}{2\pi} \int_0^{2\pi} w(x - r\cos\theta, y - r\sin\theta) d\theta.$$

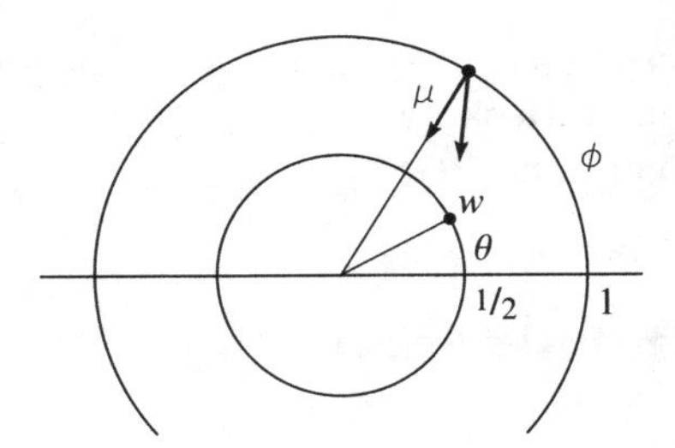

Figure 5.8: Problem 13

Show that the relationship between this average h and the gravitational inhomogeneity μ is given by the single integral

$$\mu(x, y) = 2\pi\gamma z \int_0^\infty (z^2 + r^2)^{-3/2} h(x, y, r) r \, dr.$$

13. Problem Suppose mass is distributed on a circle of radius $\frac{1}{2}$ centered on the origin with density $\rho = \rho(\theta)$. At points on the concentric circle of radius 1 in the same plane, the centrally directed component of gravitational force $\mu(\phi)$ is measured, as illustrated in Figure 5.8. Find a relationship of the form

$$\mu(\theta) = \int_0^{2\pi} k(\phi, \theta)\rho(\theta) \, d\theta$$

relating the density and centrally directed force.

14. Project Write a program that discretizes the equation in the previous problem and use it to study the direct and inverse problems. Use the program in the way 'geo' was used in some of the activities above. Your program should be capable of simulating models with random noise in the data.

5.3.3 Notes and Further Reading

For more on condition numbers and peturbation anlysis of linear systems, see, for example, G. Golub and C. van Loan, *Matrix Computations*, Johns Hopkins University Press, Baltimore, 1983. The idea of adding a small positive multiple of the identity (or some other matrix) to improve conditioning is called *regularization*. A concrete introduction to regularization, with emphasis on geophysical problems, can be found in S. Twomey, *Introduction to the Mathematics of Remote Sensing and Indirect Measurements*, Elsevier, Amsterdam, 1977 (reprinted by Dover, New York, 1996). More abstract introductions to regularization theory can be found in C. W. Groetsch, *The Theory of Tikhonov*

Regularization for Fredholm Equations of the First Kind, Pitman, Boston, 1984, and H. W. Engl, M. Hanke, and A. Neubauer, *Regularization of Inverse Problems*, Kluwer, Dordrecht, 1996.

5.4 A Whole Lotta Shakin' Goin' On

Course Level:

Linear Algebra, Differential Equations

Goal:

Investigate a simple inverse vibration problem for a binary oscillating system.

Mathematical Background:

Periodic functions, determinants, traces, eigenvalues and eigenvectors, linear differential equations

Scientific Background:

Newton's law of motion, Hooke's Law

Technology:

Calculator with matrix capability

5.4.1 Introduction

Consider an object of mass m that glides on a frictionless surface while attached to a fixed support by a stiff spring with *Hooke's constant* k (see Figure 5.9). This means that the spring resists an elongation, or compression, of x units by a force of magnitude kx. If the mass is displaced x units from its equilibrium position, then the restoring force is $-kx$, and Newton's second law gives the equation of motion

$$m\ddot{x} = -kx.$$

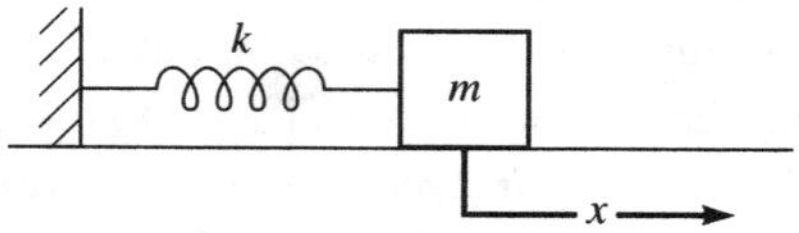

Figure 5.9: The Simplest Spring–Mass System

This simple differential equation has two linearly independent solutions, namely

$$\sin\sqrt{\frac{k}{m}}\,t; \quad \cos\sqrt{\frac{\kappa}{m}}\,t.$$

Both solutions are periodic functions with period $\sqrt{m/k}\,(2\pi)$, or, equivalently, with frequency $\omega = \sqrt{k/m}/(2\pi)$ cycles per second (= hertz). Any linear combination of these solutions is therefore also a periodic function with the same frequency, and this frequency is independent of the initial conditions. If the mass m and the "stiffness" k are known, then the *direct problem* of determining the natural frequency ω of the system has a unique solution. Solving this direct problem is the fundamental step in solving the problem of determining the system response to given initial conditions. On the other hand, the *inverse problem* of determining the mass and stiffness that give rise to an observed frequency of oscillation clearly has infinitely many solutions.

One of the aims of this module is to study the inverse problem of determining the individual masses and stiffnesses in very simple mass–spring systems, given the total mass of the components and certain frequency data. Consider the simplest case of a single known mass m confined between two springs of stiffness k_1 and k_2, respectively. We refer to such a system by using the mnemonic $(k_1, [m], k_2)$. If the natural frequency ω of this system is observed and the mass m is known, then $k_1 + k_2$ is uniquely determined (See Exercise 1) and, in fact,

$$k_1 + k_2 = m(2\pi\omega)^2.$$

If, in addition to the frequency ω of the full system, the natural frequency ω^- of the *reduced system*, obtained by disconnecting the right-hand spring (that is, the natural frequency of the system $(k_1, [m])$), is also known, then the stiffness k_1 is determined:

$$k_1 = m(2\pi\omega^-)^2.$$

Therefore, knowledge of the mass m and the two frequencies ω and ω^- of the full and reduced systems, respectively, uniquely determines the two stiffnesses.

We now show that knowledge of the sum of two masses, along with two frequency spectra, will allow us to unravel the individual masses and the stiffnesses of the springs connecting them. The ideas can be extended to systems of more than two masses, but we will treat only the simple case of a binary system.

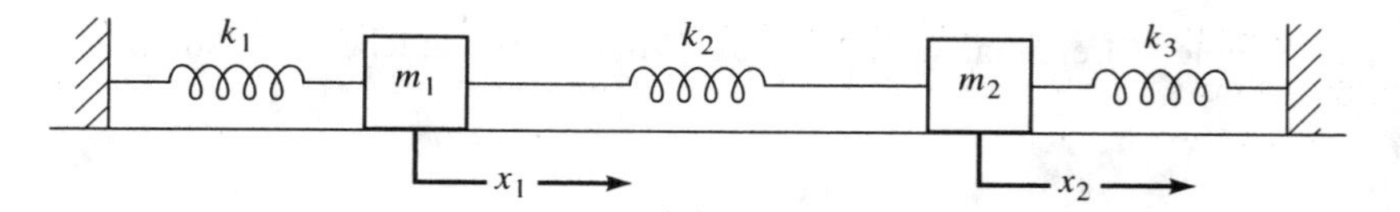

Figure 5.10: A Binary Oscillating System

Consider the system $(k_1, [m_1], k_2, [m_2], k_3)$ pictured in Figure 5.10.

If the displacements of the masses from their equilibrium positions are x_1 and x_2, respectively, then by Newton's law of motion and Hooke's Law the equations of motion are

$$m_1\ddot{x}_1 = -k_1x_1 + k_2(x_2 - x_1)$$

$$m_2\ddot{x}_2 = -k_3x_2 - k_2(x_2 - x_1),$$

or, in matrix form,

$$M\ddot{\mathbf{x}} = -K\mathbf{x},$$

where

$$M = \begin{bmatrix} m_1 & 0 \\ 0 & m_2 \end{bmatrix} \quad \text{and} \quad K = \begin{bmatrix} k_1 + k_2 & -k_2 \\ -k_2 & k_2 + k_3 \end{bmatrix}.$$

It will be convenient to convert this to an equivalent system involving a single symmetric matrix. Let

$$D = \begin{bmatrix} \sqrt{m_1} & 0 \\ 0 & \sqrt{m_2} \end{bmatrix}.$$

If we change variables by setting $\mathbf{y} = D\mathbf{x}$, then the original system becomes

$$D\ddot{\mathbf{y}} = -KD^{-1}\mathbf{y},$$

or, equivalently,

$$\ddot{\mathbf{y}} = -A\mathbf{y},$$

where

$$A = D^{-1}KD^{-1} = \begin{bmatrix} \dfrac{k_1 + k_2}{m_1} & \dfrac{-k_2}{\sqrt{m_1m_2}} \\ \dfrac{-k_2}{\sqrt{m_1m_2}} & \dfrac{k_2 + k_3}{m_2} \end{bmatrix}$$

Note that A is symmetric and positive definite. It therefore has positive eigenvalues $\lambda_1 < \lambda_2$ with associated orthonormal eigenvectors $\mathbf{u}^{(1)}$ and $\mathbf{u}^{(2)}$, respectively. The vector $\mathbf{y}(t)$ then has a unique representation of the form

$$\mathbf{y}(t) = y_1(t)\mathbf{u}^{(1)} + y_2(t)\mathbf{u}^{(2)},$$

and the differential equation (along with the orthogonality of the eigenvectors) gives

$$\ddot{y}_1 = -\lambda_1 y_1 \quad \text{and} \quad \ddot{y}_2 = -\lambda_2 y_2.$$

The solutions of these scalar differential equations are periodic functions with frequencies

$$\omega_1 = \frac{\sqrt{\lambda_1}}{2\pi} \quad \text{and} \quad \omega_2 = \frac{\sqrt{\lambda_2}}{2\pi},$$

respectively. Since $\mathbf{x} = D^{-1}\mathbf{y}$, the solutions of the original system of differential equations also have the natural frequencies ω_1 and ω_2.

The natural frequencies of the vibrating system ω_1, ω_2 are determined by the matrix A, which in turn governs the dynamics of the system. Note that A has the form

$$A = \begin{bmatrix} a & -b \\ -b & c+d \end{bmatrix},$$

where a, b, c, and d are positive. Our goal is to reconstruct A from frequency data (equivalently, eigenvalue data) and ultimately to obtain the masses and stiffnesses from our knowledge of A. Reconstructing A requires identifying four numbers: a, b, c, and d. This is clearly impossible to do if all we have to work with are the two eigenvalues of A. On the other hand, if we have (in addition to the eigenvalues of A) the eigenvalues of the *reduced* matrix

$$A^- = \begin{bmatrix} a & -b \\ -b & c \end{bmatrix},$$

then these four eigenvalues can be used to uniquely reconstruct A.

We now outline the reconstruction procedure. Suppose $\lambda_1 < \lambda_2$ and $\lambda_1^- < \lambda_2^-$ are the eigenvalues of A and A^-, respectively. Since the trace of a matrix is the sum of its eigenvalues, d is immediately determined:

$$d = trace(A) - trace(A^-) = \lambda_1 - \lambda_1^- + \lambda_2 - \lambda_2^-.$$

It remains to determine a, b, and c. Since

$$\lambda_1 \lambda_2 = \det A = ac + ad - b^2 = \lambda_1^- \lambda_2^- + ad,$$

we have

$$a = \frac{\lambda_1 \lambda_2 - \lambda_1^- \lambda_2^-}{d}.$$

And now b and c remain to be determined. Since $a+c = trace(A^-) = \lambda_1^- + \lambda_2^-$, we have

$$c = \lambda_1^- + \lambda_2^- - a,$$

and finally, from $ac - b^2 = \det A^- = \lambda_1^- \lambda_2^-$ we find that

$$b^2 = ac - \lambda_1^- \lambda_2^-,$$

which uniquely determines the positive number b. In this way the matrix A can be uniquely reconstructed from the spectral data $\{\lambda_1, \lambda_2\}$ and $\{\lambda_1^-, \lambda_2^-\}$ of the full and reduced matrices A and A^-, respectively.

Using the procedure just outlined, one can reconstruct the matrix A given the natural frequencies $\omega_1 < \omega_2$ of the full system $(k_1, [m_1], k_2, [m_2], k_3)$ and the natural frequencies $\omega_1^- < \omega_2^-$ of the reduced system $(k_1, [m_1], k_2, [m_2])$.

The four parameters a, b, c, and d of A, along with a given value for the total mass $m_2 + m_2$, can then be used to solve for the three stiffnesses k_1, k_2, and k_3 and the two masses m_1 and m_2 that comprise the vibrating system, thus solving the inverse vibration problem.

5.4.2 Activities

1. Exercise Show that the mass–spring system $(k_1, [m], k_2)$ is equivalent to the mass–spring system $(k_1 + k_2, [m])$.

2. Problem Suppose a mass m weighing 1 lb in the system $(k_1, [m], k_2)$ vibrates with a frequency $12.5/\pi$ ($\approx .398$) hertz, while the reduced system $(k_1, [m])$ has a natural frequency of $3.5/\pi$ (≈ 1.11) hertz. What are the stiffnesses of k_1 and k_2 ?

3. Problem Suppose the natural frequency of the system $(k_1, [m], k_2)$ is ω_1. If the stiffness of the first spring is doubled, the natural frequency of the resulting system $(2k_1, [m], k_2)$ is ω_2, while if the stiffness of the second spring is doubled, the system $(k_1, [m], 2k_2)$ has natural frequency ω_3. Find a condition on $\omega_1, \omega_2, \omega_3$ that guarantees that such a system exists. Show that there are infinitely many such systems.

4. Exercise Show that the two systems $(2, [2], 4, [4], 2)$ and $(1, [1], 2, [2], 1)$ have the same natural frequencies of vibration.

5. Calculation Show that the two systems $(1, [1], 1, [1], 1)$ and $((3+\sqrt{3})/2, [1], .5, [1], (3-\sqrt{3})/2)$ have the same natural frequencies of vibration.

6. Problem Show that the matrices $B = M^{-1}K$ and $A = D^{-1}KD^{-1}$ ($D = M^{1/2}$) have the same eigenvalues.

7. Problem Suppose

$$A^{-} = \begin{bmatrix} a & -b \\ -b & c \end{bmatrix},$$

where a, b, and c are positive. Show that A^{-} has distinct, real eigenvalues.

8. Problem Let

$$A = \begin{bmatrix} a & -b \\ -b & c+d \end{bmatrix} \quad \text{and} \quad A^{-} = \begin{bmatrix} a & -b \\ -b & c \end{bmatrix}.$$

Suppose the eigenvalues of A are $\lambda_1 < \lambda_2$ and the eigenvalues of A^{-} are $\lambda_1^{-} < \lambda_2^{-}$. Show that the eigenvalues *interlace*:

$$\lambda_1^{-} < \lambda_1 < \lambda_2^{-} < \lambda_2.$$

9. Problem Suppose m_1, m_2, k_1, k_2, and k_3 are positive numbers and that the quantities $m_1 + m_2$, k_2/m_2, $(k_1 + k_2)/m_1$, $(k_2 + k_3)/m_2$, and $k_2/\sqrt{m_1 m_2}$ are known. Show that m_1, m_2, k_1, k_2, and k_3 are uniquely determined.

10. Exercise Suppose the sum of two masses is 5 kg and that the vibration matrix A for the system $(k_1, [m_1], k_2, [m_2], k_3)$ is

$$\begin{bmatrix} 1 & -2/\sqrt{6} \\ -2/\sqrt{6} & 1.5 \end{bmatrix},$$

while the vibration matrix for the reduced system $(k_1, [m_1], k_2, [m_2])$ is

$$\begin{bmatrix} 1 & -2/\sqrt{6} \\ -2/\sqrt{6} & 1 \end{bmatrix}.$$

Find m_1, m_2, k_1, k_2, and k_3.

11. Problem Suppose the eigenvalues of

$$A = \begin{bmatrix} a & -b \\ -b & c+d \end{bmatrix}$$

are 2 and 7, and the eigenvalues of

$$A^{-} = \begin{bmatrix} a & -b \\ -b & c \end{bmatrix}$$

are 1 and 5. Find A and A^{-}.

12. Problem Suppose the system $(k_1, [m_1], k_2, [m_2], k_3)$ has natural frequencies $\sqrt{3}/(2\pi)$ hertz and $\sqrt{7}/(2\pi)$ hertz, while the reduced system $(k_1, [m_1], k_2, [m_2])$ has natural frequecies $1/(2\pi)$ hertz and $\sqrt{6}/(2\pi)$ hertz. If the total weight of the two objects is 2 lbs, find m_1, m_2, k_1, k_2, and k_3.

13. Calculation The total mass of the configuration $(k_1, [m_1], k_2, [m_2], k_3)$ is 5 kg. The natural frequencies of the system are .318 hertz and .450 hertz. The reduced system $(k_1, [m_1], k_2, [m_2])$ has natural frequencies of .412 hertz and .0869 hertz. Estimate the masses and stiffnesses.

In the remaining activities we investigate an inverse vibration problem involving the *location* of vibrating masses. Suppose a unit mass is attached to a tightly stretched weightless string of length one and given tension τ. The mass executes small-amplitude vertical vibrations under the influence of the tension (gravity is ignored). An exaggerated typical position is shown in Figure 5.11.

14. Problem

(a) Show that the equation of motion of the bead on the string is

$$\ddot{y} = -\tau \sin \alpha - \tau \sin \beta.$$

(b) Show that the small-amplitude assumption allows the approximations

$$\sin \alpha \approx \tan \alpha \quad \textit{and} \quad \sin \beta \approx \tan \beta,$$

which in turn lead to the *linearized* equation of motion:

$$\ddot{y} = -\frac{\tau}{x(1-x)} y.$$

15. Question In the bead-on-a-string model, are all frequencies possible (for a given tension)? What is the lowest possible frequency? To what bead position does this lowest frequency correspond?

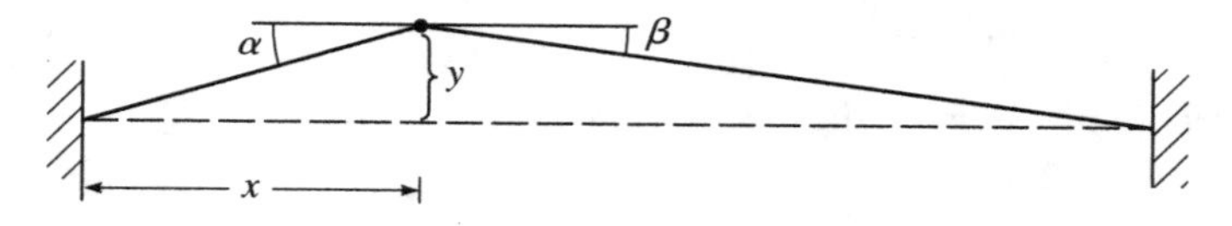

Figure 5.11: Bead on a String

16. Problem

(a) Show that the sum of the reciprocals of the distances from the bead to each support is proportional to the natural frequency.

(b) Show that (for given tension τ) the frequency of the solution of the linearized model uniquely determines the distance of the bead from the center $x = \frac{1}{2}$ of the string.

17. Problem

(a) Find the linearized equations of motion for two unit masses attached to a stretched string.

(b) Show that the natural frequencies of vibration of the linearized system uniquely determine the positions of two unit masses that are symmetrically placed with respect to the center of the string.

5.4.3 Notes and Further Reading

The main idea of this module can be extended to vibrating systems with any number of masses. For more on this, see G.M.L. Gladwell's book, *Inverse Problems in Vibration*, Martinus Nijhoff Publishers, Dordrecht, The Netherlands, 1986.

5.5 Globs and Globs

Course Level:

Linear Algebra

Goal:

Develop a discrete linear algebraic model for a projection problem.

Mathematical Background:

Multiple integration, matrix inverses

Scientific Background:

None

Technology:

Symbolic calculator, MATLAB or other high-level numerical software

5.5.1 Introduction

In this module we discuss a simple inverse problem that arises in observational astronomy: the *globular cluster* problem. A globular cluster is a far-off spherical aggregate of stars. We place the origin at the center of the cluster and take the radius of the cluster to be a positive number R. When a globular cluster is observed with a telescope, what is seen on the optical plane, or equivalently the photographic plate, of the telescope is a disk—the projection of the spherical cluster onto a two-dimensional plane. The two-dimensional density of stars on this image can then be estimated, and the challenge is to "back-project" this measured density distribution to get the three-dimensional density of stars in the cluster. We assume the density f of stars in the cluster is centrally symmetric, that is, f is a function of the distance r from the center of the cluster. The density of stars $g(x)$ on the image plane at a distance x from the center of the disk is obtained by adding all those stars on the line of projection through x perpendicular to the observational plane, as illustrated in Figure 5.12. Using the relationships $z^2 = r^2 - x^2$, we find that

$$\begin{aligned} g(x) &= 2\int_0^{\sqrt{R^2-x^2}} f(r)\, dz \\ &= 2\int_x^R \frac{rf(r)}{\sqrt{r^2-x^2}}\, dr. \end{aligned}$$

This equation is an expression of the *direct* problem of computing the projected density g given the spherical density f. The *inverse*, or back-projection, problem consists of determining the spherical density f from knowledge of the projected density g. The relationship given can be inverted by elementary

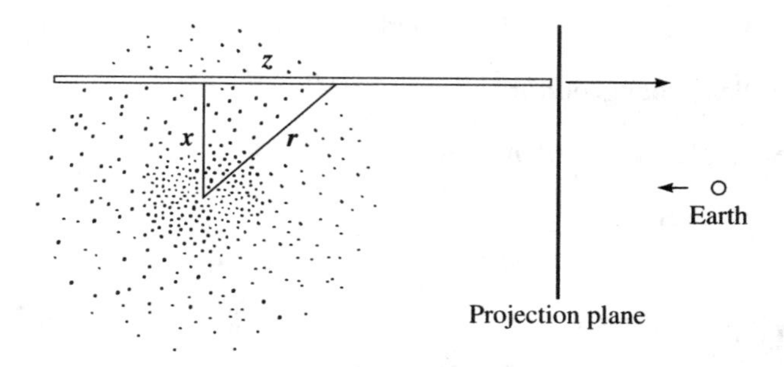

Figure 5.12: Projecting a Cluster Onto a Plane

formal manipulations (see Problem 4) to give

$$rf(r) = -\frac{1}{\pi}\frac{d}{dr}\int_r^R \frac{xg(x)}{\sqrt{x^2 - r^2}}\,dx.$$

Unfortunately, the analytical operations indicated in each of these formulas seldom lead to closed-form expressions (see Calculations 5 and 6), and therefore numerical methods are called for. We consider a very simple approximation method, of the type studied in the module *Shape Up*, in which the function f is approximated by a piecewise constant function. Specifically, for a given positive integer n, we set $h = R/n$ and $r_j = jh$, for $j = 0, 1, \ldots, n$. On each interval $[r_{j-1}, r_j)$, $j = 1, \ldots, n$, the function f will be replaced by its value at the left endpoint. That is, we make the approximation

$$f(r) \approx \sum_{j=1}^{n} f(r_{j-1})c_j(r),$$

where $c_j(r) = 1$ if $r \in [r_{j-1}, r_j)$ and $c_j(r) = 0$ otherwise. Substituting this into the forward-projection equation, we get

$$g(x) \approx \sum_{j=1}^{n} \int_x^R \frac{2r}{\sqrt{r^2 - x^2}} c_j(r)\,dr f(r_{j-1}).$$

Now if we let $x_i = ih$, $i = 0, 1, \ldots, n$, we have

$$g(x_i) \approx \sum_{j=1}^{n} \int_{ih}^R \frac{2r}{\sqrt{r^2 - i^2h^2}} c_j(r)\,dr f(r_{j-1}).$$

The calculus problem of calculating g given f has now been converted into an approximating direct linear algebra problem. If we set $y_j = f(r_{j-1})$, $j = 1, \ldots, n$ and $z_i = g(x_{i-1})$, $i = 1, \ldots, n$, then the integral transform connecting f and g is converted in the approximation to the action of a matrix A converting $\mathbf{y}$ to $\mathbf{z}$: $\mathbf{z} = A\mathbf{y}$, where

$$a_{ij} = \int_{(i-1)h}^R \frac{2r}{\sqrt{r^2 - (i-1)^2h^2}} c_j(r)\,dr.$$

From this we see that $a_{ij} = 0$ for $i > j$, since in this case $(i-1)h \geq r_j$. For $i \leq j$, we have

$$\begin{aligned} a_{ij} &= \int_{(j-1)h}^{jh} \frac{2r}{\sqrt{r^2-(i-1)^2h^2}}\,dr \\ &= 2h\left(\sqrt{j^2-(i-1)^2}-\sqrt{(j-1)^2-(i-1)^2}\right). \end{aligned}$$

The matrix A can be used on the discretized spherical density y to simulate the action of the projection of the spherical density f onto the optical plane. The program 'globd' can be used to perform such simulations. It accepts a user-defined spherical density f, a radius R, and an integer n specifying the number of subintervals in the discretization and returns an $n \times n$ matrix A and the $n \times 1$ vector $\mathbf{y}$ representing the piecewise linear approximation of f. The back-projection operation can also be simulated by using the inverse of A.

5.5.2 Activities

1. Calculation Find the projected density $g(x)$ of the spherical density $f(r) = 1 - r$ in the sphere of radius 1.

2. Exercise Show that

$$\int_r^s \frac{2x}{\sqrt{x^2-r^2}\sqrt{s^2-x^2}}\,dx = \pi.$$

(Hint: Make the change of variables $w = \sqrt{x^2-s^2}/\sqrt{r^2-s^2}$.)

3. Problem Suppose $f(r)$ is a spherical density in the sphere of radius R, and let g be the corresponding projected density. Show that

$$\int_r^R \frac{xg(x)}{\sqrt{x^2-r^2}}\,dx = \int_r^R sf(s)\int_r^s \frac{2x}{\sqrt{x^2-r^2}\sqrt{s^2-x^2}}\,dx\,ds.$$

4. Problem Use the results of Exercise 2 and Problem 3 to conclude that

$$rf(r) = -\frac{1}{\pi}\frac{d}{dr}\int_r^R \frac{xg(x)}{\sqrt{x^2-r^2}}\,dx.$$

5. Calculation Find the projected density g corresponding to the spherical density $f(r) = 1$ for $(0 \le r \le 1)$. Use the resulting projected density g in the inversion formula of Problem 4 to try to reconstruct the spherical density f.

6. Calculation Use Problem 4 to find the spherical density $f(r)$, $0 \le r \le 1$, that results in the projected density $g(x) = 1$. Then try to reconstruct g by using the spherical density f just obtained in the forward-projection formula.

7. Computation Repeat the exercise in Calculation 5, but now use the discretized version of the forward projection operator obtained from the program 'globd' (invocation: [A,y]=globd(R,n,f)) with $n = 50$ subintervals.

8. Computation Use the projected density g obtained in Calculation 1 and the discretized forward projection matrix A (with $n = 50$) obtained from the program 'globd' to approximately reconstruct the spherical density f from Problem 1. Plot the true spherical density and the reconstructed spherical density on the same axes and compare.

9. Computation Use the program 'globd' with $n = 50$ to generate a discrete version of the projected density corresponding to the spherical density $f(r) = e^{-r}$ for $0 \leq r \leq 1$. Perturb the discrete projected density with uniform random error of amplitude $ep = .001$ and use the perturbed data to reconstruct the spherical density. Repeat the procedure with $ep = .01, .1$ and plot the reconstructed spherical densities and the true spherical density and compare.

10. Computation Use $R = 1$ and the program 'globd' to find the condition numbers of the forward projection matrices for $n = 50, 100, 200$, and 400. Make a conjecture on the dependence of $cond(A)$ on n.

11. Project Develop a discretized version of the back-projection operator

$$rf(r) = -\frac{1}{\pi}\frac{d}{dr}\int_r^R \frac{xg(x)}{\sqrt{x^2 - r^2}}\,dx,$$

and test it for clean and noisy data g. Compare the accuracy and stability of your method with that resulting from the use of the inverse of the forward-projection matrix A.

12. Problem Show that the changes of variables $r^2 = R^2 - u$, $x^2 = R^2 - v$, $y(v) = g(\sqrt{R^2 - v})$, and $w(u) = f(\sqrt{R^2 - u})$ convert the equation

$$g(x) = 2\int_x^R \frac{rf(r)}{\sqrt{r^2 - x^2}}\,dr$$

into the equation

$$y(v) = \int_0^v \frac{w(u)}{\sqrt{v - u}}\,du.$$

13. Project Develop continuous and discrete inversion formulas for the forward-projection operator in Problem 12.

5.5.3 Notes and Further Reading

The transformation that takes the spherical density into the two-dimensional projected density is an example of an Abel transform. Basic information on the Abel transform and related matters can be found in P. Linz, *Analytical and Numerical Methods for Volterra Equations*, SIAM, Philadelphia, 1985. The globular cluster problem was introduced in a biological setting by S. D. Wicksell, "The corpuscle problem," *Biometrika* **17** (1925), pp. 84–99. A slightly different approach to the numerical method of this module can be found in O. H. Nestor and H. N. Olsen, "Numerical methods for reducing line and surface probe data," *SIAM Review* **2** (1960), pp. 200–207. More inverse problems in astronomy can be found in I. Craig and J. Brown, *Inverse Problems in Astronomy*, Adam Hilger, Bristol, 1986.

5.6 Tip Top

Course Level:

Linear Algebra

Goal:

Develop an estimation technique using Lagrange multipliers.

Mathematical Background:

Lagrange multipliers, positive definite matrices

Scientific Background:

Density and mass, moment of inertia, Coulomb's Law

Technology:

MATLAB or other high-level numerical software

5.6.1 Introduction

In simpler times children's games usually involved direct experience with real objects—nonvirtual reality. Tops were a favorite. The simplest top is a circular wooden disc of uniform thickness with a perpendicular central shaft, about which the top is spun with a twisting motion of the thumb and forefinger (see Figure 5.13).

In this module we use the top to illustrate the Backus–Gilbert method for estimating values of an unknown function on the basis of very limited

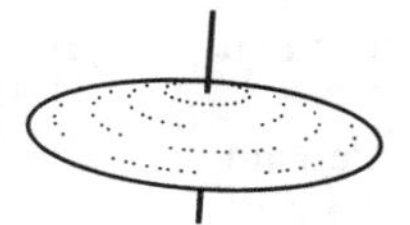

Figure 5.13: A Top

information. Suppose the disc is not uniform, but rather its mass density (say, in grams per cm^2) is a function $\rho(r)$ of the distance r from the central shaft (which is assumed to be massless). Our goal is to estimate a value of $\rho(s)$ using just a few (in fact, just two) "gross" measures of ρ. To simplify notation, we take the radius of the disc to be 1. Two overall measures of ρ come to mind. One is the total mass of the disc:

$$M = 2\pi \int_0^1 r\rho(r)\,dr.$$

If the disc is spun about its central axis with a constant angular speed ω, then its kinetic energy is $\frac{1}{2}I\omega^2$, where I, the *moment of inertia*, plays a role for rotational motion analogous to that played by mass for straight-line motion (the connection coming via the relationship between angular and tangential speed: $r\omega = v$). In terms of the distributed mass, the moment of inertia of the disc is

$$I = 2\pi \int_0^1 r^3\rho(r)\,dr.$$

Of course the *direct* problem of determining M and I, given the density ρ, is straightforward. Our interest is in the *inverse* problem of estimating the value of the density at an interior point, given M and I. Suppose we know values of M and I, or, equivalently, we know numbers μ_1 and μ_2, where

$$\mu_1 = \int_0^1 r\rho(r)\,dr \quad \text{and}$$

$$\mu_2 = \int_0^1 r^3\rho(r)\,dr.$$

On the basis of this meager information, these two "gross" measures of ρ, we would like to arrive at an estimate of the value of the density at some given radius $s \in [0, 1]$. How should we estimate $\rho(s)$? Is there really any hope?

Our only clues are in the numbers μ_1 and μ_2, so we might try to approximate $\rho(s)$ by a linear combination of the data:

$$\rho(s) \approx c_1\mu_1 + c_2\mu_2,$$

where the coefficients c_1 and c_2 depend on s. Equivalently, we have

$$\rho(s) \approx \int_0^1 (c_1 r + c_2 r^3)\rho(r)\,dr.$$

Our job is to "shape" the function $c_1 r + c_2 r^3$ (remember, c_1 and c_2 are functions of s) in a way that tends to squeeze the value $\rho(s)$ out of the integral. First, we note that substituting the constant density $\rho(r) = 1$ above leads us to impose the restriction

$$1 = \int_0^1 (c_1 r + c_2 r^3)\,dr = \frac{1}{2}c_1 + \frac{1}{4}c_2.$$

Now, imagine what would happen if we could arrange for $c_1 r + c_2 r^3$ to be concentrated around the point $r = s$. As an extreme example, consider the function

$$\Delta_s(r) = \begin{cases} 1/2\epsilon & r \in [s-\epsilon, s+\epsilon] \\ 0 & r \notin [s-\epsilon, s+\epsilon] \end{cases},$$

where ϵ is a small positive number. Notice that the constraint

$$1 = \int_0^1 \Delta_s(r)\,dr$$

is satisfied. Furthermore,

$$\int_0^1 \Delta_s(r)\rho(r)\,dr = \frac{1}{2\epsilon}\int_{s-\epsilon}^{s+\epsilon} \rho(r)\,dr \approx \rho(s).$$

This suggests that we should shape the function $c_1 r + c_2 r^3$ by penalizing values at points r that are distant from s. The effect is to concentrate the support of the function in the neighborhood of $r = s$. There are many strategies for this. For example, since the function $y(r) = |r - s|$ rewards points r that are distant from s with large functional values, one might try in some sense to *minimize*

$$|c_1 r + c_2 r^3||r - s|$$

over the span of r values. The absolute values cause some analytical irritation, so we choose to determine c_1 and c_2 so as to minimize

$$\int_0^1 (c_1 r + c_2 r^3)^2 (r - s)^2\,dr$$

subject to the constraint

$$1 = \frac{1}{2}c_1 + \frac{1}{4}c_2$$

imposed earlier.

To be definite, suppose we wish to estimate the density of the disc halfway between the center and the rim, so that $s = .5$. We then choose c_1 and c_2 to minimize

$$f(c_1, c_2) = \int_0^1 (c_1 r + c_2 r^3)^2 (r - .5)^2 \, dr$$

subject to the constraint

$$g(c_1, c_2) = .5c_1 + .25c_2 - 1 = 0.$$

This is a classic constrained quadratic minimization problem that can be handled with Lagrange multipliers. A necessary condition for a solution is the existence of a number λ satisfying

$$\nabla f(c_1, c_2) + \lambda \nabla g(c_1, c_2) = \mathbf{0}.$$

Now,

$$\begin{aligned}
\frac{\partial f}{\partial c_1} &= 2\int_0^1 (c_1 r + c_2 r^3) r (r - .5)^2 \, dr \\
&= \frac{1}{15}c_1 + \frac{11}{210}c_2 \quad \text{and} \\
\frac{\partial f}{\partial c_2} &= 2\int_0^1 (c_1 r + c_2 r^3) r^3 (r - .5)^2 \, dr \\
&= \frac{11}{2100}c_1 + \frac{11}{252}c_2,
\end{aligned}$$

and hence c_1, c_2, λ must satisfy

$$\begin{aligned}
\frac{1}{15}c_1 + \frac{11}{210}c_2 + .5\lambda &= 0, \\
\frac{11}{210}c_1 + \frac{11}{252}c_2 + .25\lambda &= 0, \quad \text{and} \\
.5c_1 + .25c_2 &= 1,
\end{aligned}$$

which has the solution $c_1 = -2/5, c_2 = 24/5$. Our estimate for $\rho(.5)$ is therefore

$$\rho(.5) \approx c_1\mu_1 + c_2\mu_2 = -\frac{2}{5}\mu_1 + \frac{24}{5}\mu_2.$$

For example, if the actual density is $\rho(r) = 1 + r$, then

$$\mu_1 = \int_0^1 r\rho(r)\,dr = \frac{5}{6}$$

and

$$\mu_2 = \int_0^1 r^3\rho(r)\,dr = \frac{9}{20}.$$

Our estimate for $\rho(.5) = 1.5$ is then

$$-\frac{2}{5}\mu_1 + \frac{24}{5}\mu_2 \approx 1.83,$$

which is in error by about 22%. While this may not be very impressive, remember that the estimate is based on only *two* meager indirect measures of information of the unknown function ρ.

We now consider the estimation problem in a bit more generality. Suppose we have n linear measures of an unknown function ρ given by

$$\mu_j = \int_0^1 g_j(r)\rho(r)\,dr, \quad j = 1, 2, \ldots, n,$$

where g_j are known functions (in the example treated above, $g_1(r) = r$ and $g_2(r) = r^3$; we take the integration interval $[0, 1]$ for convenience). Our goal is to estimate ρ at s by a linear combination of the measurements:

$$\begin{aligned}\rho(s) &\approx c_1(s)\mu_1 + \cdots + c_n(s)\mu_n \\ &= \int_0^1 A(s,r)\rho(r)\,dr,\end{aligned}$$

where

$$A(s,r) = \sum_1^n c_j(s)g_j(r).$$

As in the example of the top, we insist that $A(s, r)$ perform an "averaging" in the sense that

$$\int_0^1 A(s,r)\,dr = 1,$$

and we "shape" the function $A(s, r)$ by penalizing points r distant from s by requiring that

$$\int_0^1 (A(s,r)(r-s))^2\,dr$$

is a minimum.

To summarize: We wish to find a vector

$$\mathbf{c}(s) = [c_1(s), c_2(s), \ldots, c_n(s)]^T \in R^n$$

that satisfies

$$1 = \sum_{j=1}^{n} c_j(s) \int_0^1 g_j(r)\,dr = (\mathbf{c}(s), \boldsymbol{\gamma}),$$

where $(\cdot,\cdot)$ is the usual euclidean inner product on R^n, and $\boldsymbol{\gamma} \in R^n$ is the vector with components

$$\gamma_j = \int_0^1 g_j(r)\,dr, \quad j = 1, 2, \ldots, n.$$

Furthermore, we minimize the quadratic form

$$\begin{aligned}
\int_0^1 (A(s,r)(s-r))^2\,dr &= \int_0^1 \left(\sum_{j=1}^{n} c_j(s)g_j(r)\right)\left(\sum_{i=1}^{n} c_i(s)g_i(r)\right)(s-r)^2\,dr \\
&= \sum_{i=1}^{n}\sum_{j=1}^{n} \int_0^1 g_i(r)g_j(r)(s-r)^2\,dr c_j(s)c_i(s) \\
&= (G(s)\mathbf{c}(s), \mathbf{c}(s)),
\end{aligned}$$

where the $n \times n$ symmetric matrix $G(s)$ has entries

$$G(s)_{i,j} = \int_0^1 g_i(r)g_j(r)(s-r)^2\,dr, \quad i, j = 1, \ldots, n.$$

The procedure may be succinctly summarized as

$$\text{Minimize:} \quad (G(s)\mathbf{c}(s), \mathbf{c}(s))$$

$$\text{Subject to:} \quad (\mathbf{c}(s), \boldsymbol{\gamma}) = 1.$$

Routine calculations give $\nabla g(\mathbf{c}) = \boldsymbol{\gamma}$ and $\nabla f(\mathbf{c}) = 2G\mathbf{c}$, and therefore

$$\begin{aligned}
\nabla f(\mathbf{c}) + \lambda \nabla g(\mathbf{c}) &= \mathbf{0} \\
(\mathbf{c}, \boldsymbol{\gamma}) &= 1
\end{aligned}$$

gives

$$2G\mathbf{c} + \lambda\boldsymbol{\gamma} = \mathbf{0},$$

and hence

$$\mathbf{c} = -\frac{\lambda}{2}G^{-1}\boldsymbol{\gamma}.$$

Substituting this into the constraint gives

$$\mathbf{c} = G^{-1}\boldsymbol{\gamma}/(G^{-1}\boldsymbol{\gamma}, \boldsymbol{\gamma}).$$

Our estimate then has the following simple explicit form:

$$\rho(s) \approx (\mathbf{c}(s), \boldsymbol{\mu}) = (G^{-1}(s)\boldsymbol{\gamma}, \boldsymbol{\mu})/(G^{-1}(s)\boldsymbol{\gamma}, \boldsymbol{\gamma}).$$

5.6.2 Activities

1. Problem Show that for each fixed s, the form

$$\langle f, g\rangle_s = \int_0^1 f(r)g(r)(s-r)^2\,dr$$

is an *inner product* on the space of continuous functions on $[0, 1]$. Note that the entries of the matrix $G(s)$ discussed in the introduction may then be expressed as

$$G(s)_{i,j} = \langle g_i, g_j\rangle_s.$$

2. Problem Show that if the functions $g_1, \ldots, g_n$ are continuous and linearly independent, then the matrix $G(s)$ is positive definite (and hence invertible). (Hint: Consider $\langle T_x, T_x\rangle_s$, where T is the transformation defined by $T_x = x_1 g_1 + \cdots + x_n g_n$.)

3. Problem Give a geometrical interpretation of the following conditions: $(G\mathbf{c}, \mathbf{c})$ a minimum, while $(\mathbf{c}, \boldsymbol{\gamma}) = 1$, where G is a positive definite matrix and γ is a fixed vector.

4. Exercise Suppose a circular disc of radius 1 cm has total mass $\frac{2}{3}$ gm and its moment of inertia about the central axis is .3 gm(cm)^2. Assuming that the density $\rho(r)$ (gm/cm^2) is a function of the radius r alone, estimate $\rho(.5)$.

5. Computation Calculate the matrix $G(s)$ generated by the functions $g_i(r) = r^{i-1}$ for $0 \le r \le 1$ and $i = 1, 2, \ldots, n$, corresponding to the "moments" of an unknown function. Find the condition number of $G(.5)$ for $n = 5, 10, 20,$ and 40.

6. Problem Consider a solid ball of radius 1 with density distribution $\rho(r)$ (gm/cm^3, say) that is a function of the distance r from the center. Find functions g_1 and g_2 so that

$$\mu_1 = \int_0^1 \rho(r) g_1(r)\, dr$$

is the total mass of the ball, and

$$\mu_2 = \int_0^1 \rho(r) g_2(r)\, dr$$

is the moment of inertia of the ball about a central axis. Find the matrix $G(s)$ used to estimate the density $\rho(s)$. If, in appropriate units, the mass is .83 and the moment of inertia is .47, estimate $\rho(.25)$.

7. Problem Consider two parallel line segments, each of length 2, that are 1 unit apart, say, the segment $x \in [-1, 1]$ situated one unit above the segment $z \in [-1, 1]$. Suppose electric charge with density $f(z)$ is laid out on the lower segment. According to Coulomb's Law, the electric field strength induced by a given charge on another unit charge is proportional to the charge and inversely proportional to the square of the distance between the charges. Suppose, for simplicity, that we take the constant of proportionality to be 1. Show that this leads to the following relationship between the induced electric field $E(x)$ at a point $x \in [-1, 1]$ on the upper segment and the charge density $f(z)$ on the lower segment:

$$E(x) = \int_{-1}^1 \frac{f(z)}{(x-z)^2 + 1}\, dz.$$

8. Calculation Referring to Problem 7, suppose that the electric field (in appropriate units) at the endpoints of the upper segment is measured as $E(-1) = E(1) = 1.12$. Use the method of this module to estimate the charge density at the center of the lower segment, $f(0)$.

9. Problem The temperature $u(x)$ at time 1 and position $x \in (-\infty, \infty)$ on an infinite bar is related to the initial temperature distribution $u_0(z)$ by

$$u(x) = \frac{1}{2\sqrt{\pi}} \int_{-\infty}^{\infty} u_0(z) \exp\left(-\frac{(x-z)^2}{4}\right) dz$$

(we have taken certain constants to be 1 for convenience). Suppose measured temperatures $u(x_1), u(x_2), u(x_3)$ at time 1 are available at certain positions

x_1, x_2, x_3. Identify symbolically the matrix G and the vector γ that the method of this module uses to estimate the initial temperature $u_0(0)$ at the origin.

5.6.3 Notes and Further Reading

Accounts of the Lagrange multiplier method for constrained minimization can be found in many calculus texts, for example R. Courant, *Differential and Integral Calculus*, Vol. 2, Interscience, New York, 1967. The Backus–Gilbert method was developed for geophysical data analysis and has been generalized in many directions. The original paper (G. E. Backus and J. F. Gilbert, "The resolving power of gross earth data," *Geophysical Journal of the Royal Astronomical Society* **16** (1968), pp. 169–205) is quite readable. As indicated in Problem 6, direct application of the method, even for relatively small numbers of measurements, can lead to severely ill-conditioned systems. Such systems require special numerical techniques (see P. C. Hansen, *Rank Deficient and Discrete Ill-Posed Problems: Numerical Aspects of Linear Inversion*, SIAM, Philadelphia, 1998). A derivation of the integral equation in Problem 9 can be found, for example, in H. Pollard, *Applied Mathematics: An Introduction*, Addison-Wesley, Reading, MA, 1972.

A

Selected Answers & Advice

A.1 Inverse Problems in Precalculus

A.1.1 A Little Squirt

1. Question Eliminating the parameter shows that the curve is a parabola.

2. Exercise $T = \sqrt{2h/g}$. The descent time depends only on the vertical velocity and the vertical distance fallen, both of which are independent of D.

4. Problem $R = 2\sqrt{h(D-h)}$

5. Question Neither; the range is independent of gravity.

6. Problem $0 \le R \le D$.

7. Calculation The curve is a semi-ellipse:

$$\frac{R^2}{144} + \frac{(h-6)^2}{36} = 1 \quad R \ge 0.$$

8. Problem $h = D/2 \pm \sqrt{D^2 - R^2}/2$. At the higher hole, the drop has less horizontal velocity but a longer distance (and hence a longer time) to fall, giving the same horizontal range as the drop through the lower hole, which has a larger horizontal velocity but less time to fall.

9. Question The height of the hole is unique if and only if the depth is equal to the range. In this case the height of the hole is half the depth.

10. Problem $h = \dfrac{R_1^2 - R_2^2}{4(D_1 - D_2)}$.

11. Problem Assuming no air resistance, the horizontal velocity is *always* positive, so the trajectory is never perpendicular to the horizontal. (The calculus approach is to show that dy/dx is always finite).

A.1.2 A Cheap Shot

1. Exercise $T = 2v_y/g$. Note that the flight time is independent of the horizontal velocity.

3. Problem For any given $R > 0$, and any given $v_x > 0$, the choice $v_y = gR/2v_x$ gives the required range R.

4. Calculation

(a) The curve is one branch of a hyperbola.

(b) If $v_y = 400$ and $R = 5{,}000$, then $v_x = 201.25$.

(c) For $v_x = 300$, $v_y \approx 268.3$.

(d) $\pi/4$ rad.

6. Question From Problem 5 we see that the maximum range occurs at $\pi/4$ rad and is equal to v^2/g.

7. Problem

$$y_{\max} = \frac{v^2 \sin^2 \theta}{2g}.$$

8. Problem The minimum muzzle velocity $v_{\min} = \sqrt{gR}$ occurs at $\theta = \pi/4$.

9. Problem If $0 \leq R^* \leq v^2/2g$, then the range R^* is attained for $\theta_1 = \frac{1}{2} \arcsin gR^*/v^2$. The other solution is $\theta_2 = (\pi/2) - \theta_1$.

10. Problem $T = 2v(\sin \theta - \tan \alpha \cos \theta)$.

11. Problem $R = x(T) = 2(v_x v_y - v_x^2 \tan \alpha)$.

12. Calculation The nearest point to the origin appears to be about $(.84, 3.31)$. Therefore,

$$\theta \approx \arctan 3.31/.84 \approx 75.8 \text{ degrees}.$$

That is, the angle roughly bisects the angle between the battlefield and the vertical.

13. Problem By Problem 11,

$$R = \frac{2v^2}{g}(\sin\theta\cos\theta - \cos^2\theta\tan\alpha).$$

14. Problem For fixed v, maximizing R is equivalent to maximizing

$$2\sin\theta\cos\theta - 2\cos^2\theta\tan\alpha = (\sin(2\theta - \alpha) - \sin\alpha)/\cos\alpha.$$

This has a maximum when $2\theta - \alpha = \pi/2$, that is,

$$\theta - \alpha = \pi/2 - \theta;$$

i.e., the barrel bisects the angle between the inclined plane and the vertical (as hinted at in Problem 12).

15. Problem For fixed α, the equation

$$\frac{gR}{v^2}\cos\alpha + \sin\alpha = \sin(2\theta - \alpha)$$

has exactly two solutions in θ.

A.1.3 das Rheingold

1. Question No. Any two sources with the same mass placed symmetrically with respect to the measurement site will give the same measurement.

2. Question The curve is a parabola with vertex (x, G) and focus

$$\left(x, G + \frac{1}{2\sqrt{G}}\right).$$

3. Question Changes in (x, G) reposition the vertex. As G increases, the vertex rises and the parabola "tightens." The length of the latus rectum is $1/G$.

5. Problem

(a) $M > G \Rightarrow s = x \pm \sqrt{(M - G)/G}$.

(b) $M = G \Rightarrow s = x$.

(c) $M < G$ is impossible.

6. Exercise

(a) If $M > G$, then two objects symmetrically placed with respect to the vertical through the observation site could account for the same observed vertical force.

(b) If $M = G$, then the vertical force is the actual force, so the source lies directly below the observation site.

(c) $M < G$ is impossible since the vertical component of force cannot be greater than the force itself.

7. Question The source position is midway between the observation sites.

8. Exercise $(\sqrt{2}, 3)$ is the unique source.

9. Question No.

10. Exercise $\{(1,2), (3,10)\}$.

11. Calculation The source curves appear to intersect at about $(2.4, 7.2)$ and $(8.7, 160)$, indicating two possible sources.

12. Calculation The trace function indicates that the source nearest the origin is about $(1.86, 4.22)$.

13. Calculation Possible sources are about $(.69, 9.2)$ and $(2.4, 29.5)$.

14. Calculation No. The source curves do not intersect.

15. Question Observations at distinct sites generate source curves that are parabolas with distinct vertices. Any pair of such parabolas may intersect at most at two points.

16. Problem Let $A = G_1G_2(x_1 - x_2)^2$ and $B = (G_1 - G_2)^2$. If $A < B$, the observations are inconsistent. If $A > B$, the observations arise from two distinct sources. Otherwise the observation is accounted for by a unique source.

18. Problem Suppose (s, M) is a source giving rise to the observations (x_1, G_1) and (x_2, G_2), with $x_1 < s < x_2$. If (s_1, M_1) is another source leading to these two observations, then one can show that if $s < s_1 < x_2$, then $M - G_1 > G_1(s - x_1)^2$, and hence (s, M) cannot lead to the observation (x_1, G_1). A similar contradiction is reached if one assumes $x_1 < s_1 < s$.

19. Exercise $s = 3$ and $M = 10$.

20. Problem

(a) Make the change of variables $x \to x - x_0$ and $G \to G/G_0$.

(b) A symbolic calculator eases the algebraic pain.

23. Problem Make two observations, say, (x_1, G_1) and (x_2, G_2), where $x_1 < x_2$. If $G_1 = G_2$, then $s = (x_1 + x_2)/2$ (Question 7). Suppose now that $G_1 < G_2$ (the other case is similar). Then there is an $x_3 > x_2$ such that $G_3 < G_1$. One is then guaranteed that $x_1 < s < x_3$. (Why?)

24. Problem First, find x_1 and x_2 so that $x_1 < s < x_2$ (Problem 23). Let $x_3 = (x_1 + x_2)/2$. If $G_2 > G_1$, then $x_3 < s < x_2$; otherwise, $x_1 < s < x_3$. Continue in this way halving the interval in which the source is trapped at each step.

A.1.4 Splish Splash

1. Question Galileo's law of falling bodies.

2. Question The velocity of the falling stone is *variable* (it depends on time), while the velocity of sound is *constant*.

3. Exercise 3.2 seconds.

4. Exercise 0.15 seconds.

5. Problem $t = \sqrt{2d/g} + d/c$.

6. Problem The relevant equation

$$d^2 - 2(ct + c^2/g)\,d + (ct)^2 = 0$$

has two roots:

$$ct + c^2/g \pm \sqrt{2c^3t/g + c^4/g^2}.$$

Clearly $d < ct$ (remember t is *total* time), so the root corresponding to the "+" is physically impossible.

7. Exercise $d \approx 77$m.

8. Problem By Problem 5, $t > \sqrt{2d/g}$, i.e., $d < gt^2/2$. Physically, this is a consequence of Galileo's law of falling bodies, since the fall time is less than the total time of return of the signal.

10. Problem By Galileo's law of falling bodies, $\delta = gt^2/2 - d > 0$. By Problem 6,

$$\delta = \frac{1}{2}gt^2 - d = \frac{1}{2}gt^2 - ct - \frac{c^2}{g} + \frac{c^2}{g}\sqrt{1 + 2gt/c}.$$

Using the approximation for $\sqrt{1+x}$ from Calculation 6 then gives

$$\delta \approx \frac{1}{2}\frac{g^2t^3}{c}.$$

11. Question From Exercise 7, $t = 4.2$ and hence $2gt/c \approx 0.25$, therefore the approximation (as indicated in Calculation 9) should be reasonably good.

12. Exercise The approximation gives $d \approx 76$.

A.1.5 Snookered

1. Question Four.

2. Question Three (assuming one can strike the cue without disturbing the target).

3. Exercise $(11/30, 1), (17/35, 0), (0, 34/45), (1, 36/55)$.

4. Problem

$$\left(\frac{x_1(1-y_2)+x_2(1-y_1)}{2-(y_1+y_2)}, 1\right), \left(\frac{x_1y_2+x_2y_1}{y_1+y_2}, 0\right),$$

$$\left(0, \frac{y_1x_2+y_2x_1}{x_1+x_2}\right), \left(1, \frac{y_1(1-x_2)+y_2(1-x_1)}{2-(x_1+x_2)}\right).$$

5. Problem No. The "bank" point would have to be $(.55, 1)$, but the line through $(.55, 1)$ and $(.8, .5)$ does not contain $(1, 0)$.

6. Problem Any point on the line containing $(.42, 1)$ and having slope $= 2.5$ will do.

7. Question It seems that θ and $\psi(\theta)$ determine the slope of the target line; the additional information $r(\theta)$ gives a point on the target line.

8. Question No, but it seems that its orientation, i.e., its slope, can be determined.

9. Exercise slope $= -\tan(\pi/24)$, $\left(\frac{10\sqrt{3}}{1+\sqrt{3}}, \frac{10}{1+\sqrt{3}}\right)$ is on the line.

10. Problem Two tangent lines determine the circle.

A.1.6 Goethe's Gondoliers

1. Question 100 m/sec.

2. Question No, the depth–velocity hyperbolas for the two observations do not intersect.

3. Problem Drop a vertical line from the geophone to the point $2d$ units below the surface. The shortest distance from the source to this point is along a straight-line segment.

4. Exercise $d = 31$ m and $v = 100$ m/sec.

5. Calculation $d \approx 50$ m and $v \approx 100$ m/sec.

6. Problem $T_1X_2 > T_2X_1$.

7. Problem $T \to 2d/v$ as $X \to 0$. Note that d/v is the time for the *vertical* signal to reach the surface from the bedrock. For X large, $T \approx X/v$, which is the time for *horizontal* travel.

8. Problem The singers are 962.5 feet apart. The listener is 825 feet from A.

9. Problem

(a) P.

(b) No.

(c) Somewhere on the perpendicular bisector of PQ.

(d) No.

(e) No, the strike is at one of two points of intersection of two circles.

10. Problem $x^2/(2.2022)^2 - y^2/(4.4945)^2 = 1$ unit of 10^3 ft.

11. Problem If the observer is nearer to the right vertex of the hyperbola, then the strike is near $(2429, \pm 2099)$ (units in feet). Otherwise the strike is near $(-2429, \pm 2099)$.

A.2 Inverse Problems in Calculus

A.2.1 Strange Salami

1. Exercise $C(x) = \dfrac{x(2x+3)}{3(x+2)}$.

2. Exercise $C(x) = ((x+1)\ln(x+1) - x)/x$.

3. Problem For any $x \in [0, 1]$,

$$\left| \int_0^x f_n(u)\,du - \int_0^x f(u)\,du \right| \le a_n x \le a_n \to 0;$$

therefore,

$$\int_0^x f_n(u)\,du \to \int_0^x f(u)\,du > 0.$$

Similarly,

$$\int_0^x u f_n(u)\,du \to \int_0^x u f(u)\,du,$$

and hence

$$C_n(x) \to C(x)$$

for all x.

4. Calculation The centroid of the left quarter is about 0.1145; that of the right quarter is about .8853.

5. Problem Since $uf(u) > 0$ for $u > 0$,

$$0 < \frac{\int_0^x uf(u)\,du}{\int_0^x f(u)\,du} < \frac{x\int_0^x f(u)\,du}{\int_0^x f(u)\,du} = x.$$

The centroid of $[0, x]$ lies *within* the segment $[0, x]$.

6. Problem $0 < C(x) < x \Rightarrow \lim_{x\to 0^+} C(x) = 0$.

7. Question Clearly, $C(x)$ should be increasing. As the bar grows to the right, its centroid should shift to the right.

8. Problem The quotient rule gives

$$C'(x) = \frac{xf(x)\int_0^x f(u)\,du - \int_0^x uf(u)\,du f(x)}{(\int_0^x f(u)\,du)^2} = f(x)(x - C(x))/\int_0^x f(u)\,du.$$

All of the factors in the last expression are positive.

9. Problem By Problem 6, C can be continuously extended to $[0, 1]$ by setting $C(0) = 0$. Then

$$C'(x) = \lim_{x\to 0^+} \frac{C(x) - C(0)}{x - 0}.$$

Since $0 < C(x)/x < 1$, if this limit exists, then

$$0 \le C'(0) \le 1.$$

If $f(0) \neq 0$, then

$$\begin{aligned} C'(0) &= \lim_{x\to 0^+} \frac{xf(x)}{\int_0^x f(u)\,du + xf(x)} \\ &= \lim_{x\to 0^+} \frac{f(x)}{(\int_0^x f(u)\,du)/x + f(x)} = \frac{1}{2}. \end{aligned}$$

10. Exercise $\lim_{x\to 0^+} f(x) = \infty$, so we would be unable to construct such a bar. However, $C(x) = 2x/5$.

11. Question For a given $x > 0$, measure a length x from the left end and cut. Now use the dull edge of the knife to balance this cut-off segment. Mark the balance point and use the tape measure to find $C(x)$.

12. Question Yes. Multiplying a given density by a positive, non-1 constant gives a different density with the same centroid function.

13. Problem

(a) Equate the centroid functions, cross-multiply, differentiate, and rearrange.

14. Problem Integrate the result of Problem 13.

16. Calculation $f(u) = 2(u+1)/3$.

17. Exercise

(c) By Problem 8(a):

$$f_n(1) = C_n'(1)/(1 - C_n(1)) = \frac{2n^2 + n}{n - 2} \to \infty.$$

18. Problem Try a function of the form $f(u) = Cu^a$. $f(u) = 2u/3$ works.

A.2.2 Shape Up!

1. Problem $f(y) = \text{constant} \times y^{1/4}$.

2. Question The effective shape has units sec/ft since $F(y) = T'(y)$.

4. Problem If $f(y) = r$, a constant, then $T(y) = 200r^2\sqrt{y}$.

7. Problem Since

$$T''(y) = 200f(y)(f'(y)y - \tfrac{1}{4}f(y))y^{-3/2},$$

one might try to find an f so that the factor in parentheses has exactly two roots in $(0, 1)$. For example, one might try a quadratic which is positive on $(0, 1)$ and for which the roots of the parenthetical factor are $\frac{1}{4}$ and $\frac{3}{4}$. A little experimentation shows that

$$f(y) = -\tfrac{1}{7}y^2 + \tfrac{1}{3}y + \tfrac{3}{16}$$

works. Of course, infinitely many other possibilities exist.

8. Problem ∞.

9. Problem $|T(y) - \tilde{T}(y)| \leq 400\epsilon\sqrt{y}$.

12. Problem $T_n(y) \to y/2$, and the shape corresponding to this drain time function is $f(y) = y^{1/4}/(10\sqrt{2})$. However,

$$T_n'(1/2) = 1/2 + 3\sqrt{n}/2 \to \infty.$$

Therefore, $f_n(1/2) \to \infty$.

15. Problem Since

$$dV/dt = -a\sqrt{2gy},$$

if dV/dy were a constant, then y would be constant; i.e., no water would drain from the tank. For the cylinder in question,

$$\sqrt{y} = -\frac{2 \times 10^{-3}}{\pi}t + \sqrt{.4}.$$

For $t \in [0, 3]$ the values of dV/dt differ from $dV/dt(1.5)$ by less than 2 percent.

A.2.3 What Goes Around Comes Around

1. Exercise $\ddot{\mathbf{r}} = -a^2\mathbf{r}$, so the force is central. Since $\|\dot{\mathbf{r}}\| = ar$, we have

$$\|\ddot{\mathbf{r}}\| = \|\dot{\mathbf{r}}\|^2/r.$$

2. Exercise Direct calculation.

3. Problem

$$\begin{aligned}\frac{d}{dt}\left(\frac{\mathbf{r}}{r}\right) &= \frac{(\mathbf{r}\cdot\mathbf{r})\dot{\mathbf{r}} - (\dot{\mathbf{r}}\cdot\mathbf{r})\mathbf{r}}{r^3} \\ &= \frac{(\mathbf{r}\times\dot{\mathbf{r}})\times\mathbf{r}}{r^3}\end{aligned}$$

by Exercise 2.

5. Question No. Just about any example works, e.g., $\mathbf{r} = t\mathbf{i} + e^t\mathbf{j}$.

6. Problem Draw a line through the center of force parallel to the line of motion. Consider the areas of the triangles swept out by the radial segment over two consecutive time intervals of the same duration.

7. Exercise The second derivative of a linear function vanishes.

8. Exercise Differentiate $\frac{1}{2}r^2\dot{\theta} = \text{constant}$.

9. Problem If the force is central, then it is planar and hence may be parameterized in the form

$$\dot{\mathbf{r}} = r\cos\theta\mathbf{i} + r\sin\theta\mathbf{j}.$$

Compute $\ddot{\mathbf{r}}$ and use the result of Exercise 8.

10. Problem If the orbit is represented by $x = r\cos\theta$, $y = r\sin\theta$, then

$$x\dot{y} - y\dot{x} = r^2\dot{\theta} = \text{constant}.$$

Therefore, $x\ddot{y} - y\ddot{x} = 0$, and hence

$$\mathbf{r} \times \ddot{\mathbf{r}} = (x\mathbf{i} + y\mathbf{j}) \times (\ddot{x}\mathbf{i} + \ddot{y}\mathbf{j}) = \mathbf{0};$$

i.e., the force is centrally directed.

11. Problem $r\dot{\theta}^2 = r(hu^2)^2 = h^2u^3$.

12. Problem Since $r = u^{-1}$, Problem 11 gives

$$\dot{r} = -u^{-2}\frac{du}{d\theta}\dot{\theta} = -u^{-2}\frac{du}{d\theta}(hu^2) = -h\frac{du}{d\theta}.$$

Now differentiate again.

14. Problem Here, $u = 1/r = A\csc\theta$ where $A = 1/a$. This leads to

$$\frac{d^2u}{d\theta^2} + u = 2A\csc^3\theta \propto u^3,$$

and hence the force is proportional to $u^2u^3 = r^{-5}$.

15. Problem If $r = a/(1 + e\cos\theta)$, then $u = A + B\cos\theta$ and hence $(d^2u/d\theta^2) + u$ is a constant.

Therefore an inverse-square force acts.

16. Problem Inverse cube.

17. Problem The force has the form $A/r^5 + B/r^3$.

18. Problem Here, $abu = (b^2\cos^2\theta + a^2\sin^2\theta)^{1/2}$, where a and b are the semiaxes of the ellipse. Routine, but tedious, calculations give $(d^2u/d\theta^2) + u \propto 1/u^3$. So the force is proportional to r.

19. Problem The force is *repulsive* and proportional to r.

20. Problem The force is proportional to a linear combination of $1/r^3$ and $1/r^7$.

A.2.4 Hanging Out

1. Question No. A vertically translated shape could result from the same weight distribution. One could fix the shape by specifying the y-intercept.

2. Problem Integrate the basic equation from 0 to 1, interchange the order of integration, etc.

3. Problem Differentiate the basic equation.

4. Problem Integrate the basic equation from 0 to x to get $y(x)$, then compute $y(-x)$.

5. Problem 5 $\int_0^x w(u)\,du = H\frac{d}{dx}y(x) = H\frac{d}{dx}y(-x) = -\int_0^{-x} w(u)\,du.$ Now differentiate with respect to x.

6. Exercise $y = x^4$ and $y = 3x^4 + 7$ are both solutions.

8. Problem Integrate the basic equation from 0 to x and use Problem 2.

12. Exercise $w(x) = \frac{\pi}{4}\cos\frac{\pi x}{2}$.

13. Problem Use Problem 11.

15. Problem Use Problem 11.

16. Calculation $w(u) = 1/4$ for $u < 0$; $w(u) = 3u^2/2 + 1/4$ for $u \geq 0$.

17. Problem Use Problem 8.

22. Problem $w = aH$.

23. Calculation The minimum occurs at $x_0 = 1/a$ where $x_0 = \coth(x_0)$. The unique positive solution of this equation satisfies $\sinh(x_0) > 1.5$).

25. Problem $f(x) = sx + 1 - \coth x$ is continuous, vanishes at 0, and has a positive derivative for positive x sufficiently small. Hence $f(x_1)$ is positive for some $x_1 > 0$. But $f(x) \to -\infty$ as $x \to \infty$, so $f(x)$ has a positive root. Since 0 is a root, were there another positive root, $f''(x)$ would vanish for some positive x. But f'' is strictly negative.

26. Calculation Approximately, $r(.1) = .20$, $r(.5) = .93$, $r(1) = 1.6$, $r(2) = 2.5$, $r(4) = 3.4$, $r(10) = 4.5$.

28. Problem About .34 lbs.

29. Problem $a\cosh x/a = a + (x^2/2a) + (\sinh\theta)(x^3/6a^3)$.

30. Problem For $-1 \leq x \leq 1$, the error is bounded by $C/(6a^3)$.

A.2.5 Two Will Get You Three

2. Question The vertex is at $(0, h)$ and the focus at $(0, h - (g/8v^2))$.

4. Exercise If there is no resistance, then $\ddot{x} = 0$ and hence $\ddot{y} = 2av^2$, a constant.

5. Problem The position vector is $\mathbf{r} = x\mathbf{i} + ((-g/2v^2)x^2 + 1)\mathbf{j}$. Equating components in $\ddot{\mathbf{r}} = -f(r, \dot{r})\dot{\mathbf{r}}$ leads to $\dot{x} = v$ and hence $\ddot{x} = -f(r, \dot{r})v = 0$, that is, $f(r, \dot{r}) = 0$.

6. Problem $g(y)$ is proportional to y^3.

7. Problem $g(y)$ is proportional to y^{-3}.

A.2.6 Uncommonly Interesting

1. Exercise $r(t) = \dfrac{1}{2t + 10} + .02.$

4. Calculation The inflection point occurs at $t \approx 1.42$.

6. Problem $r_n(1/n) = ne/(1 + e)^2$.

7. Problem $r_k = \dfrac{\ln u(t_k + h) - \ln u(t_k)}{h} = \dfrac{\dfrac{u'(t_k)h}{u(t_k)} + (\ln u)''(\theta)\dfrac{h^2}{2}}{h} < r(t_k).$

A.3 Inverse Problems in Differential Equations

A.3.1 Stirred, Not Shaken

1. Exercise $a \approx .02$ and $r \approx 105$ gallons/day.

2. Problem Let c_i be the concentration at time t_i. Then $c_1/c_2 = \gamma(R)$ where $R = r/V$ and

$$\gamma(R) = (1 - e^{-t_1 R})/(1 - e^{-t_2 R}).$$

Since $\gamma'(R) > 0$, γ is invertible. Therefore, R, and hence r/V, is uniquely determined by $R = \gamma^{-1}(c_1/c_2)$. a is now uniquely determined by $c_1 = a(1 - e^{-rt_1/V})$.

4. Question No, the best one can do is identify a and the ratio r/V.

7. Problem Assuming c has a bounded third derivative,

$$c'(t_i) - \frac{c(t_{i+1}) - c(t_{i-1})}{2h} = O(h^2),$$

and therefore

$$a_i - a(t_i) = \frac{r}{V}\left(\frac{c(t_{i+1}) - c(t_{i-1})}{2h} - \frac{c_{i+1} - c_i}{2h}\right) + c(t_i) - c_i + O(h^2).$$

9. Problem The concentration at time t satisfies

$$c(t) - c_0 = (a - c_0)\left(1 - \left(\frac{V_0}{V}\right)^{\frac{r}{r-\rho}}\right),$$

where $V(t) = (r - \rho)t + V_0$. We then have

$$\frac{c_2 - c_0}{c_1 - c_0} = \frac{1 - (V_0/V_2)^{\frac{r}{r-\rho}}}{1 - (V_0/V_1)^{\frac{r}{r-\rho}}} := \gamma(x),$$

where $x = r/(r - \rho) > 1$. But for $0 < a < b < 1$ and $x > 1$, $\gamma'(x) > 0$, and hence γ is invertible. Therefore, x is uniquely determined from the given information. Since $r - \rho$ is uniquely determined by V_2, t_2, it follows that r and ρ are uniquely determined. Finally, a is uniquely determined by substituting t_1 into the formula for $c(t)$.

10. Calculation Inflow rate $\approx$11 gallons/hour; outflow rate $\approx$2 gallons/hour; inflow concentration $\approx$ 4 percent.

A.3.2 Slip Sliding Away

1. Problem Suppose the particle starts at point (a, b) on the circle $x^2 + (y - r)^2 = r^2$, and let $s(t)$ be the distance from the origin to the point's position at time t. Then $s = -(gb/2\sqrt{a^2 + b^2})t^2 + \sqrt{a^2 + b^2}$, and hence the time of descent is $T = \sqrt{2(a^2 + b^2)/(gb)} = 2\sqrt{r/g}$.

5. Problem If $\ddot{s} = g \sin\alpha - k\dot{s}$ and $\dot{s}(0) = 0$, then

$$\dot{s} = \frac{g \sin\alpha}{k}(1 - e^{-kt}),$$

and hence

$$s = \frac{g}{k}\left(t - \frac{1 - e^{-kt}}{k}\right)\sin\alpha.$$

6. Problem $\dot{D}_k = \frac{g}{k}(1 - e^{-kt}) > 0$, and hence $\dot{D}_k(t) \to 0$ as $t \to 0^+$ and $\dot{D}_k(t) \to g/k$ as $t \to \infty$.

8. Problem Substituting $s = D(t)\sin\alpha$ into the equation of motion and setting $\sigma = \sin\alpha$, we find that $f(z\sigma)/\sigma$ is independent of σ for all possible velocities z and all $\sigma \in (0, 1]$. For any $w \in (0, z]$, we then have (letting $\sigma = w/z$)

$$\frac{f(z)}{z} = \frac{1}{z}\frac{f(\sigma z)}{\sigma} = \frac{f(w)}{w}.$$

That is, $f(z)/z = k$ a constant.

9. Problem The solution of the equations of motion is

$$s = \frac{g}{k}\left(1 - \cos\sqrt{k}t\right)\sin\alpha.$$

12. Problem $f(s, \dot{s}) = s + 2\dot{s}$.

16. Problem The curves have the form $s = D(t)g\sin\alpha$ where

$$D(t) = \left(1 - e^{-\frac{a}{2}t}\left(1 + \frac{a}{2}t\right)\right)/b \quad \text{if} \quad a^2 = 4b,$$

$$D(t) = e^{-\frac{a}{2}t}\left(\frac{a}{2\beta b} - \frac{1}{b}\cos\beta t\right) + \frac{1}{b} \quad \text{if} \quad a^2 < 4b, \quad \text{and}$$

$$D(t) = \frac{r_2}{r_1 - r_2}e^{r_1 t} - \left(\frac{r_2 g}{r_1 - r_2} + 1\right) r_2 e^{r_2 t} + 1 \quad \text{if} \quad a^2 > 4b,$$

where $\beta = \sqrt{4b - a^2}/2$ and

$$r_{1,2} = \frac{-a \pm \sqrt{a^2 - 4b}}{2}.$$

In each case the equitemporal curves are circles.

18. Calculation $a \approx 2.48876, \theta \approx 1.78157$.

19. Problem $f(\theta) = (1 - \cos\theta)/(\theta - \sin\theta)$ is strictly decreasing on $(0, \pi]$ and has range $(2/\pi, \infty)$. So, if $y/x \in (2/\pi, \infty)$, there is exactly one $\theta \in (0, \pi]$ with

$$\frac{y}{x} = \frac{1 - \cos\theta}{1 - \sin\theta}.$$

Let $a = x/(1 - \sin\theta)$; then

$$x = a(1 - \sin\theta) \quad y = a(1 - \cos\theta).$$

20. Problem Let s be the length of arc descended in time t. By conservation of energy, $ds/dt = \sqrt{2gy}$. However,

$$ds = a\sqrt{2}\sqrt{1 - \cos\theta}d\theta.$$

Therefore, $\sqrt{a}d\theta = \sqrt{g}dt$, and hence $\theta = \sqrt{g/a}t$. For a given time t the particle reaches position

$$\left(a\left(\sqrt{g/a}t - \sin\sqrt{g/a}t\right), a\left(1 - \cos\sqrt{g/a}t\right)\right).$$

The set of such points parameterized by $a > 0$ is the equitemporal curve.

A.3.3 It's a Drag

3. Problem $f'(0) = cd > 1$ and $f(0) = 0$, so $\lim_{x\to 0^+}(f(x)/x) > 1$.

4. Problem $f(c) = c(1 - e^{-dc}) < c$. Since $f(s) > s$ for all small positive s, f has a fixed point in $(0, c)$.

5. Problem If f has two positive fixed points, then, since 0 is a fixed point, f will have *three* fixed points. Then $f''(x)$ must vanish at some positive x. However, $f''(x) < 0$.

6. Problem Suppose $f(x) > x$ for some $x > p$. Since $f(x) \to c$ as $x \to \infty$, f has a fixed point q with $p < x < q$. But f has a unique positive fixed point.

7. Problem For $x > 0$, $e^x > 1 + x$ and hence $x < (x + 1)(1 - e^{-x})$. Using this with $x = cd - 1$, we get $f((cd - 1)/d) > (cd - 1)/d$, and hence, by Problem 6, $(cd - 1)/d < p$.

18. Problem Set $R'(\theta) = 0$ and substitute $(a \sec\theta)R(\theta) = 1 - e^{-A(\theta)R(\theta)}$.

28. Problem By Exercise 27,

$$e^{eh/w} = e^{(1/2w)+1-(w/2)} < e/w.$$

if and only if

$$w < e^{(w/2)-(1/2w)}$$

Letting $j(w)$ be this last expression, we have $j(1) = 1 = j'(1)$ and $j''(w) > 0$. Therefore, by Problem 25,

$$w < e^{(w/2)-(1/2w)}; \quad \text{hence} \quad e^{eh/w} < e/w;$$

therefore, $\sin\theta < 1/\sqrt{2}$, by Problem 26.

A.3.4 Ups and Downs

2. Problem $k = 49$ N/m and $c \approx 13.1$ N sec/m.

3. Problem $k = 140/3$ N/m and $c \approx 0.06$ N sec/m.

4. Problem $c = 2/(t_2 - t_1)$ and $k = 1/(t_2 - t_1)^2$.

5. Exercise $c = 0.2$ N sec/m and $k = 0.01$ N/m.

6. Computation $c \approx 0.2$ and $k \approx 0.01$.

7. Computation $c \approx 0.064$ and $k \approx 4.8$.

8. Computation $c \approx 3$ and $k \approx 2$.

10. Computation $c/m \approx 1.9$ and $k/m \approx 0.97$.

A.3.5 A Hot Time

2. Exercise $A = 22$, $u(0) = 84.5$ and $\alpha \approx 0.045$.

3. Problem $\dfrac{u_3 - u_1}{u_2 - u_1} = \dfrac{1 - E^{\gamma}}{1 - E} =: f(E)$, where $\gamma = (t_1 - t_3)/(t_1 - t_2) > 1$ and $E = e^{\alpha(t_1 - t_2)} < 1$. But f is one-to-one.

4. Problem If $t_k - t_0 = k(t_1 - t_0)$, then $A - u(t_k) = (A - u(t_{k-1}))e^{\alpha(t_1 - t_0)}$.

12. Problem $b(x) = 1$ and $b_n(x) = (\pi^2 - n)/\pi^2$.

A.3.6 Weird Weirs

1. Exercise $r(h) = 32\sqrt{(h - .5)^3}/3$ for $h \geq .5$ and $r(h) = 0$ otherwise.

2. Problem $r(h) = 2\pi h^2$.

3. Problem $f(y) = \sqrt{y}/(2\pi)$.

4. Problem $f(y) = 15y/64$.

5. Problem $f(y)$ proportional to $1/\sqrt{y}$. No.

6. Problem $\left|\dfrac{r(h) - r(0)}{h}\right| \leq 16\sqrt{h}M$.

10. Problem Plug the result of Problem 9 into the left-hand side in Problem 7, reverse the order of integration, and use the result of Problem 6.

12. Problem Integrate by parts.

14. Problem $|r_f(h) - r_g(h)| \leq 32\sqrt{H^3}\epsilon/3$.

A.4 Inverse Problems in Linear Algebra

A.4.1 Cause and Identity

1. Problem No.

2. Problem Yes. Diodes, 1.82; Transistors, 2.91; Resistors, no charge.

3. Question No.

4. Problem $A = \text{diag}(\sqrt{(1+\epsilon)\epsilon}, \sqrt{\epsilon/(1+\epsilon)})$.

9. Problem Use Problems 6, 7, and 8.

11. Exercise $b \notin R(A)$, but $A^T A$ is invertible. The unique least-squares solution is $[0, 3/2]^T$.

12. Exercise $\frac{1}{3}[1, 1]^T$.

16. Exercise $A = [1, 3, 2; -1, 0, 1]$.

17. Problem The columns of A represent benzene, acetylene, and naphthalene, respectively.

18. Problem Heating oil: 3 of type 1 to 7 of type 2. Gasoline: 8 of type 1 to 2 of type 2.

20. Problem $A = BX^T(XX^T)^{-1}$.

21. Problem

(c) $\Gamma = [1, -R_1; -1/R_2, R_1/R_2 + 1]$, $L = [1 + R_1/R_2, -R_1; -1/R_2, 1]$ where R_1 is the "series" resistor and R_2 is the "shunt" resistor.

(d) For a Γ circuit (with $i_1 \neq 0$), v_2 determines R_1 and i_2 determines R_2. For an L circuit (with $v_1 \neq 0$), i_2 determines R_2 and v_1 determines R_1.

A.4.2 L'ART Pour L'Art

1. Exercise No, not generally. The matrix connecting the individual scores to the given averages is not invertible.

2. Problem Yes, except for the square.

3. Exercise Yes.

5. Exercise $[0.428, 0.715]$.

7. Problem For example, $[1, 0, .5, 0, 1, 0, .5, 0, 1]$ and $[1.2241, 0, 0.2759, 0, 1, 0, 0.7241, 0, 0.7759]$.

8. Exercise

$$x = [0, 0, 0, 0, 1, 1, 1, 0, 0].$$

$$V = [111000000; 010001000; 001001001; 000100010].$$

A.4.3 Nonpolitical Pull

4. Problem $\mu(x) = x/\sqrt{1+x^2} + (1-x)/\sqrt{1+(1-x)^2}$.

8. Calculation

$$\mu(x) = \ln\left(\sqrt{x^2+1} - x\right) / \left(\sqrt{(x-1)^2+1} - (x-1)\right) + x\sqrt{(x-1)^2+1} - (x-1)\sqrt{x^2+1}.$$

10. Calculation The scroll feature of your calculator will get a workout.

13. Problem $k(\phi,\theta) = 4G\dfrac{2-\cos(\phi-\theta)}{(5-4\cos(\phi-\theta))^{3/2}}$.

A.4.4 A Whole Lotta Shakin' Goin' On

1. Exercise The equation of motion of the system $(k_1, [m], k_2)$ is the same as that of the system $(k_1 + k_2, [m])$.

2. Problem $k_1 = 49$ lbs/ft and $k_2 = 576$ lbs/ft.

3. Problem $\omega_3^2 - 3\omega_1^2 + \omega_2^2 = 0$ is a necessary and sufficient condition.

4. Exercise The system $(2, [2], 4, [4], 2)$ has the same equation of motion as the system $(1, [1], 2, [2], 1)$.

5. Calculation $\lambda_1 = 1$ and $\lambda_2 = 3$.

8. Problem The characteristic polynomials of A^- and A are $p_{A^-}(\lambda) = (\lambda - a)(\lambda - c) - b^2$ and $p_A(\lambda) = p_{A^-}(\lambda) + (a - \lambda)d$.

$$\begin{aligned}
p_{A^-}(a) &= -b^2, & p_{A^-}(\infty) &> 0 \Rightarrow \lambda_1^- < a < \lambda_2^- \\
p_A(\lambda_1^-) &= (a - \lambda_1^-)d > 0, & p_A(a) &= -b^2 < 0 \Rightarrow \lambda_1^- < \lambda_1 < a < \lambda_2 \\
p_A(\lambda_2^-) &= (a - \lambda_2^-)d < 0, & p_A(\infty) &> 0 \Rightarrow \lambda_1 < \lambda_2^- < \lambda_2
\end{aligned}$$

9. Problem

$$\begin{aligned}
k_3/m_2 \text{ and } (k_2+k_3)/m_2 \text{ known} &\quad\Rightarrow\quad k_2/m_2 \text{ known} \\
k_2/\sqrt{m_1 m_2} \text{ known and } k_2/m_2 \text{ known} &\quad\Rightarrow\quad \sqrt{m_1/m_2} \text{ known} \\
m_1 + m_2 \text{ known and } \sqrt{m_1/m_2} \text{ known} &\quad\Rightarrow\quad m_1, m_2 \text{ known}
\end{aligned}$$

Finally, if $k_1 + k_2, k_2 + k_3, k_3$ are known, then the stiffnesses are uniquely determined.

10. Exercise $m_1 = 3, \quad m_2 = 2, \quad k_1 = k_3 = 1, \quad k_2 = 2.$

11. Problem $A = [3, -2; -2, 6]; \quad A^- = [3, -2; -2, 3].$

12. Problem $(3, [1], 2, [1], 3).$

13. Calculation $\approx (2, [1], 4, [4], 20).$

15. Question No, the frequencies are bounded below, but not above. The lowest frequency occurs when the object is at the midpoint of the string.

16. Problem

(b) $|.5 - x| = \sqrt{1/4 - \lambda^2/\tau}.$

17. Problem

(b) Suppose the masses are x units and $1 - x$ units, respectively, from the left end of the string. Then the equations of motion are $\ddot{y} = -Ay$ where $A = [a + b, -b; -b, a + b]$ with $a = 1/x$ and $b = 1/(1 - 2x)$. Then $\sigma(A) = \{a, a + 2b\}$, and hence the spectrum determines a and b. Since

$$|\lambda_2 - \lambda_1| = \frac{2}{1 - 2x},$$

the difference of the frequencies determines x.

A.4.5 Globs and Globs

1. Calculation $g(x) = -x^2 \ln\left(1 + \sqrt{1 - x^2}\right) - x^2 \ln x - \sqrt{1 - x^2}.$

5. Calculation $g(x) = 2\sqrt{1 - x^2}.$

6. Calculation $f(r) = \dfrac{1}{\pi\sqrt{1 - r^2}}.$

A.4.6 Tip Top

2. Problem Define $T : R^n \to C$ by $T\mathbf{x} = x_1 g_1 + \cdots + x_n g_n$. Then $\mathbf{x}^T G\mathbf{x} = \langle T\mathbf{x}, T\mathbf{x}\rangle_s \geq 0$ and $\mathbf{x}^T G\mathbf{x} = 0$ implies $\|T\mathbf{x}\|_s = 0$, but then $x_1 = \cdots = x_n = 0$, since the g_i are linearly independent.

3. Problem $E = \{\mathbf{c} : (G\mathbf{c}, \mathbf{c}) = \text{const.}\}$ is an ellipsoid centered on the origin. $\{\mathbf{c} : (\mathbf{c}, \boldsymbol{\lambda}) = 1\}$ is a hyperplane. The two conditions define an ellipsoid in the hyperplane. The minimum is achieved when this ellipsoid in the hyperplane shrinks to a point, i.e., when the hyperplane is tangent to the ellipsoid.

4. Exercise $\rho(.5) \approx 0.24.$

6. Problem $\rho(.25) \approx 0.22.$

B

MATLAB Scripts

B.1 MATLAB Scripts

This appendix contains scripts, in the author's heavily FORTRAN-accented MATLAB, that may be used in some of the *Computations* in this book. The source codes may also be downloaded from the author's Webpage (http://math.uc.edu).

B.2 Contents

```
Chapter 3      Calculus

Module 3.1     cent        density from centroid data
Module 3.2     shape       shape from drain times
Module 3.2     drain       drain times from shape
Module 3.3     orbit0,2    planar orbits
Module 3.6     rate        interest rate from balance history

Chapter 4      Differential Equations

Module 4.3     shot        range of a given shot
Module 4.3     range       range function
Module 4.3     theta       elevation angles from range
Module 4.3     thopt       optimal angle of elevation
Module 4.4     coeff       damping and stiffness coefficients
Module 4.4     coeff1      coefficients from discrete inputs
Module 4.5     temps       discrete temperature values
```

```
Module 4.5      coeff2    coefficient in heat flow
Module 4.6      flow      direct problem for weir notch
Module 4.6      notch     inverse problem for weir notch

Chapter 5       Linear Algebra

Module 5.2      proj      projection onto a hyperplane
Module 5.2      art1      algebraic reconstruction technique
Module 5.2      displa    screen display in elementary tomography
Module 5.2      shade     shading routine for tomographic display
Module 5.3      geo       estimate of distributed mass
Module 5.5      globd     discretized stellar tomography
```

B.3 Calculus Scripts

B.3.1 Module 3.1

```
%..............................................................
%  Reconstruction of one-dimensional density from centroid data
%..............................................................
%
        function [f,t] = cent(n,c,ep);
%
%..............................................................
%
%  Inputs:
%
% n  = number of subintervals + 1
% ep = noise level in centroid data (set ep=0 for "clean" data)
% c  = vector of centroid data
%
%  Outputs:
%
% t  = vector of abscissas (t(i)=(i-1)/n)
% f  = vector of computed density values
%
%..............................................................
%
%  The centroid function, c, is related to the density function, f, by:
%  c(x)integral(0 to x)(f(u)du) = integral(0 to x)(uf(u)du)
%  Trapezoid  rule is used on integrals and the resulting
%  equation is solved for the  f  value at the right hand
%  end point in terms of previously obtained f values.
%
%..............................................................
%
%    Initializations
%
  h=1/(n-1); f(1)=1; t(1)=0.; t(2)=h;
%
```

```
%    Corrupt centroid data with random noise
%
  cep=c'+ep*(ones(n,1)-2*rand(n,1));
%
%    Approximate density values
%
  f(2)=cep(2)/(h-cep(2));
%
for k=3:n;
        sum=0.;
        for i=2:k-1;
            sum=sum+(cep(k)-(i-1)*h)*f(i);
        end;
        sum=2*sum+cep(k);
        f(k)=sum/(k*h-h-cep(k));
        t(k)=(k-1)*h;
end;
```

B.3.2 Module 3.2

```
%..............................................................
% Program to approximate shape from drain time data
%..............................................................
%
        function [f]=shape(N,T,ep);
%
%..............................................................
%
% Inputs:
%
% N  = number of subintervals of [0,1]
% T  = vector of drain times (N+1) components
% ep = amplitude of random noise to be added to T
%
% Outputs:
%
% f = vector of computed shape values at heights j/n, j=0,...,N
%
%..............................................................
%
% This program computes values for the shape function by a simple
% finite difference method, given error corrupted values of drain
% times for various water depths.
%
%..............................................................
%
% Pollute T with random noise
%
        T=T+ep*(ones(1,N+1)-2*rand(1,N+1));
```

```
%
% Generate shape approximations
%
        h=1/N;
        for j=1:N;
                F=(T(j+1)-T(j))/h;
                f(j)=sqrt(sqrt(j*h)*F)/10;
        end;
        f(N+1)=f(N);


%...................................................................
% Program to compute drain time function
%...................................................................
%
        function [y,T]=drain(N,f_,ep);
%
%...................................................................
%
% Inputs:
%
% f_ = shape function
% N  = number of subintervals of [0,1]
% ep = amplitude of uniform random error in data
%
% Outputs:
%
% T = vector of drain times (N+1 components)
% y = vector of depths
%
%...................................................................
%
% This program uses the recursive method to generate drain times from
% error corrupted values of the shape function.
%
%...................................................................
%
% Generate random noise vector
%
        noise=ep*(ones(N+1,1)-2*rand(N+1,1));
%
% Initializations
%
        H=1/sqrt(N);
        T(1)=0; y(1)=0;
        f(1)=feval(f_,0)+noise(1);
%
% Generate approximate drain times
%
```

```
        for j=2:N+1;
                y(j)=(j-1)/N;
                f(j)=feval(f_,(j-1)/N)+noise(j);
                T(j)=T(j-1)+f(j-1)^2*(sqrt(j)-sqrt(j-1))*H*200;
        end;
```

B.3.3 Module 3.3

```
%...................................................................
% Orbit Problem with constant force
%...................................................................
%
        function zdot=orbit0(t,z);
%
%...................................................................
%
% Sets up second order differential equation
%
% (d^2/dt^2)r = - r/|r|
%
% as a system of four first order differential equations for passing
% to MATLAB ode routines.
% Vector z =[x,y,xdot,ydot]
%
%...................................................................
zdot(1)=z(3);
zdot(2)=z(4);
zdot(3)=-z(1)/(z(1)^2+z(2)^2);
zdot(4)=-z(2)/(z(1)^2+z(2)^2);


%...................................................................
% Orbit Problem with inverse-square force
%...................................................................
%
% Sets up second order vector differential equation
%
%  (d^2/dt^2)r = - r/|r|^3
%
%  as a system of four first order equations for passing to
%  MATLAB ode routines.
%  Vector z=[x,y,xdot,ydot]
%...................................................................
%
function zdot=orbit2(t,z);
zdot(1)=z(3);
zdot(2)=z(4);
zdot(3)=-z(1)/sqrt(z(1)^2+z(2)^2)^3;
zdot(4)=-z(2)/sqrt(z(1)^2+z(2)^2)^3;
```

B.3.4 Module 3.6

```
%...........................................................
% Computation of variable interest rate from corrupted balan
%...........................................................
%
      function [rint,rder,t]=rate(n,term,u,r0,ep);
%
%................................................................
%
% Inputs:
%
% n    = number of periods in term
% term = term of investment in years
% u    = vector of balance history (u(1)=initial balance)
% r0   = initial value of interest rate
% ep   = amplitude of error in balance data
%
% Outputs:
%
% rint = reconstructed rate using integral method
% rder = reconstructed rate using derivative method
%    t = n-vector of times at which rate is computed
%
%................................................................
%
% This program reconstructs variable interest rates using both the
% derivative and the integral method
%
%................................................................
%
% Initializations
%
      h=term/(n-1); rint(1)=r0; rder(1)=r0; t(1)=0;
%
% Corrupt data with amplitude ep uniform random error
%
      uep = u' + ep*(ones(n,1)-2*rand(n,1));
%
% Approximate rate by both methods
%
      for k=1:n-1;
              t(k+1)=k*h;
              rint(k+1)=(2*(uep(k+1)-uep(k))/h - rint(k)*uep(k))/uep(k+1);
              rder(k+1)=(log(uep(k+1))-log(uep(k)))/h;
      end;
      rder(n)=rder(n-1);
```

B.4 Differential Equation Scripts

B.4.1 Module 4.3

```
%..................................................................
% Program that finds the range of a given shot with linear resistance
%..................................................................
%
        function r=shot(k,v,theta,ro);
%
%..................................................................
%
%  Inputs:
%
%  k     = coefficient of resistance
%  v     = muzzle velocity (ft/sec)
%  theta = angle of elevation (radians)
%  ro    = initial approximation for range
%
%  Outputs:
%
%  r     = range of shot (ft)
%
%..................................................................
%
%  Program uses fixed point iteration on the equation
%
%  R(theta)=v cos(theta)(1-exp(-A(theta) R(theta))/k
%
%..................................................................
%
%    Check if there is resistance
%
     if k==0,
        r=v*v/32.2*sin(2*theta);
     else
%    compute range for this theta by fixed point iteration
     A=k*(k*tan(theta)/32.2 + 1/(v*cos(theta)));
     rn=v*cos(theta)*(1-exp(-A*ro))/k;
     while(abs(ro-rn)>2*eps*rn);
          ro=rn;
          rn=v*cos(theta)*(1-exp(-A*ro))/k;
     end;
     r=rn;
     end;

%..................................................................
% Projectile Range Function in a Linearly Resisting Medium
%..................................................................
%
        function [t,r]=range(k,v,n);
```

```
%
%......................................................................
%
% Inputs:
%
% k = coefficient of resistance
% v = muzzle velocity (ft/sec)
% n = number of angles -1 at which range is computed
%
% Outputs:
%
% t = vector of elevation angles (radians)
% r = vector of corresponding ranges (ft)
%
%......................................................................
%
% Program uses the program 'shot' to compute the range (ft)
% for equally spaced angles in [0,pi/2].
%......................................................................
%
%    Initializations
%
     h=pi/(2*n);
     t(1)=0.; r(1)=0.;
%
%    generation of points (t,r)
%
for i=1:n;
     ti=t(i)+h;
     t(i+1)=ti;
%
%    compute range for this ti
%
     if i==1 ro=.001; else ro=r(i); end;
     r(i+1)=shot(k,v,ti,ro);
end;


%......................................................................
% Program computes elevation angles for Tartaglia's inverse problem
%......................................................................
%
        function [th1,th2]=theta(k,v,r,th10,th20)
%
%......................................................................
%
% Inputs:
%
%   r    = given suboptimal range
%   k    = resistance constant
```

```
%   v    = muzzle velocity (ft/sec)
%   th10 = initial approximation of smaller angle
%   th20 = initial approximation of larger angle
%
%  Outputs:
%
%   th1  = smaller angle of elevation
%   th2  = larger angle of elevation
%
%..........................................................................
%
% Newton's method is used to compute the two roots of the equation
%
%  1/exp(a*t+b*sqrt(t^2-r^2))-1+a*t=0, where a=k/v and b=k^2/g
%
%  then, theta = arcsec(t/r)
%
%..............................................................................
%
%  Initializations
%
        g=32.2; a=k/v; b=k^2/g;
%
% Compute first angle
%
        t0=r/cos(th10);
        num=1/exp(a*t0+b*sqrt(t0^2-r^2))-1+a*t0;
        dnom=a-(a+b*t0/sqrt(t0^2-r^2))/exp(a*t0+b*sqrt(t0^2-r^2));
        tn=t0-num/dnom;
        while abs(t0-tn)>2*eps*abs(t0);
                t0=tn;
                num=1/exp(a*t0+b*sqrt(t0^2-r^2))-1+a*t0;
                dnom=a-(a+b*t0/sqrt(t0^2-r^2))/exp(a*t0+b*sqrt(t0^2-r^2));
                tn=t0-num/dnom;
        end;
        th1=acos(r/tn);
%
% Compute second angle
%
        t0=r/cos(th20);
        num=1/exp(a*t0+b*sqrt(t0^2-r^2))-1+a*t0;
        dnom=a-(a+b*t0/sqrt(t0^2-r^2))/exp(a*t0+b*sqrt(t0^2-r^2));
        tn=t0-num/dnom;
        while abs(t0-tn)>2*eps*abs(t0);
                t0=tn;
                num=1/exp(a*t0+b*sqrt(t0^2-r^2))-1+a*t0;
                dnom=a-(a+b*t0/sqrt(t0^2-r^2))/exp(a*t0+b*sqrt(t0^2-r^2));
                tn=t0-num/dnom;
         end;
        th2=acos(r/tn);
```

```
%.................................................................
% Finds the optimal angle of elevation in a linearly resisting medium
%.................................................................
%
        function theta=thopt(k,v);
%
%.................................................................
%
%  Inputs:
%
%   k = coefficient of resistance (k>0)
%   v = muzzle velocity (ft/sec)
%
%  Outputs:
%
%   theta = angle of elevation leading to maximum range
%
%.................................................................
%
% Newton's method is used to find the root of
%
%        x-exp(h x)=0, where
%        x=(e s)/(s + c),  s=sin(theta), c=v k/g
%
%.................................................................
%
% Initializations
%
        c=v*k/32.2;  s0=pi/4; e=exp(1);
        x0=e*s0/(s0+c);  h=(1-c^2)/e;
%
%  Compute new approximation
%
        xn=x0-(x0-exp(h*x0))/(1-h*exp(h*x0));
%
        while abs(xn-x0)>2*eps*x0
                x0=xn;
                xn=x0-(x0-exp(h*x0))/(1-h*exp(h*x0));
        end;
        theta=asin(c*xn/(e-xn));
```

B.4.2 Module 4.4

```
%.....................................................................
% Routine to estimate coefficients in a 1-D dynamical model
%.....................................................................
%
        function [x1,x2]=coeff(y,T,N,ep);
%
%.....................................................................
%
%  Inputs:
%
% y     = input function of observed states
% T     = length of time of observation
% N     = number of time subintervals for approximation
% ep    = amplitude of random noise in state vector
%
%  Outputs:
%
% x1(1)= damping coefficient by least squares finite differences
% x1(2)= stiffness coefficient by least squares finite differences
% x2(1)= damping coefficient by finite element method
% x2(2)= stiffness coefficient by finite element method
%.....................................................................
%
% Estimates the damping coefficient c and the stiffness coefficient k
% for the differential equation
%                          y''+cy'+ky = 0
% by two methods: finite difference approximation + least squares, and
% finite elements + least squares
%
%.....................................................................
h=T/N;
%---   Form vector of perturbed state data ---
for i=1:N+1;
        yy(i)=feval(y,(i-1)*h)*(1+ep*(1-2*rand(1,1)));
end;
% --- Form rhs vector and matrix  ---
for i=2:N;
        b(i-1)=(-yy(i+1)+2*yy(i)-yy(i-1))/h^2;
        A(i-1,1)=(yy(i+1)-yy(i-1))/(2*h);
        A(i-1,2)=yy(i);
        B(i-1,1)=A(i-1,1);
        B(i-1,2)=(yy(i-1)+4*yy(i)+yy(i+1))/6;
end;
%   --- Compute coefficient estimates   ---
x1=A\b';  x2=B\b';
```

```
%..........................................................................
% Routine to estimate coefficients in a 1-D dynamical model
% Input is a measured vector of state values
%..........................................................................
%
        function [x]=coeff1(yy,N,h);
%
%..........................................................................
%
%  Inputs:
%
%  yy    = input vector of observed states
%  h     = time between state observations
%  N     = number of observations
%
%  Outputs:
%
%  x(1)= damping coefficient by least squares finite differences
%  x(2)= stiffness coefficient by least squares finite differences
%..........................................................................
%
% Estimates the damping coefficient c and the stiffness coefficient k
% for the differential equation
%                       y''+cy'+ky = 0
% by  finite difference approximation + least squares
%
%..........................................................................
%
% --- Form rhs vector and matrix  ---
%
for i=2:N-1;
        b(i-1)=(-yy(i+1)+2*yy(i)-yy(i-1))/h^2;
        A(i-1,1)=(yy(i+1)-yy(i-1))/(2*h);
        A(i-1,2)=yy(i);
end;
%
%   --- Compute coefficient estimates   ---
%
        x=A\b';
```

B.4.3 Module 4.5

```
%..............................................................
% Program to generate discrete temperature values
%..............................................................
%
      function [U,g] = temps(u,n,ep);
%
%..............................................................
%
%  Inputs:
%
%       u   = u(x,t):user supplied temperature distribution function
%       n   = number of subintervals desired in space discretization
%       ep  = error amplitude in simulated data
%
%  Outputs:
%
%       U is a (n-1)x5 matrix of temperatures at Gauss times of
%          interior mesh positions
%
%       g is a vector of initial temperature data
%
%..............................................................
%
%       U(i,j) = temperatue at time t(j) and position x(i)
%       t(j)   = jth Gauss-Laguerre node, j=1,...,5
%       x(i)   = i*h, i=1, ..., n-1
%
%..............................................................
%
% --------- 5 point Gauss-Laguerre nodes    ------------------
%
t(1)=.263560319718; t(2)=1.413403059107; t(3)=3.596425771041;
t(4)=7.085810005859; t(5)=12.640800844276;
%-----------------------------------------------------------
h=1/n;
for i=1:n-1;
        for j=1:5;
        U(i,j)=feval(u,i*h,t(j))*(1+ep*(1-2*rand));
        end;
        g(i)=feval(u,i*h,0)*(1+ep*(1-2*rand));
end;

%..............................................................
% Approximation of coefficient in heat problem
%..............................................................
%
      function [b,x]=coeff2(U,g,n);
%
```

```
%.................................................................
%
%  Inputs:
%
%       U is an (n-1)X5 matrix of sampled temperatures at interior
%          mesh points and gauss times (U(i,j)=temp. at position i*h,
%          time t(j) = jth gauss time (see 'temps'))
%
%       g is an (n-1) vector of initial temperatures at interior points
%
%  Outputs:
%       b = vector of approximate coefficient values, b(i) approximates
%           coefficient at position x(i)
%       x = vector of interior nodes, x(i)=i*h
%
%.................................................................
%
% This program estimates the space-dependent coefficient b(x) in the
% heat problem
%
%       b(x)(d/dt)u = (d^2/dx^2)u   , 0<x<1, 0<t
%
%       u(0,t)=0, u(1,t)=0, u(x,0)=g(x)
%
%.................................................................
%
h=1/n;
%
% --------- 5 point Gauss-Laguerre weights  -------------------
%
w(1)=.521755610583; w(2)=.398666811083; w(3)=.0759424496817;
w(4)=.00361175867992; w(5)=.0000233699723858;
%-----------------------------------------------------------
%
for i=1:n-1;
        x(i)=i*h;
        v(i)=U(i,:)*w';
        ww(i)= v(i)-g(i);
end;
b(1)=(v(2)-2*v(1))/(ww(1)*h^2);
b(n-1)= (v(n-2)-2*v(n-1))/(ww(n-1)*h^2);
for i=2:n-2;
        b(i)=(v(i+1)-2*v(i)+v(i-1))/(ww(i)*h^2);
end;
```

B.4.4 Module 4.6

```
%.............................................................
% Direct problem for weir notch
%.............................................................
%
function [r,h]=flow(f_,H,N,ep);
%
%.............................................................
%
% Inputs:
%
% f_  = supplied notch shape function
% H   = maximum height of water
% N   = number of subintervals for discretization of rate function
% ep  = amplitude of relative random error in data
%
% Outputs:
%
% r   = function value for flow rate
% h   = abscissas for flow function
%
%.............................................................
%
delh=H/N;
for i=1:N
        h(i)=(i-1)*delh;
        fvec(i)=(1+ep*(1-2*rand(1,1)))*feval(f_,(i-1)*delh);
        for j=1:i;
                A(i,j)=16*sqrt(i-j+1)*delh^(3/2);
        end;
end;
r=A*fvec';

%.............................................................
% Inverse problem for weir notch
%.............................................................
%
function [f,y]=notch(r_,H,N,ep);
%
%.............................................................
%
% Inputs:
%
% r_  = supplied desired rate function
% H   = maximum height of water
% N   = number of subintervals for discretization of rate function
% ep  = amplitude of relative random error in data
%
```

```
% Outputs:
%
% f   = function value for notch shape
% y   = abscissas for notch shape
%
%.................................................................
%
h=H/N;
for i=1:N
        y(i)=(i-1)*h;
        rvec(i)=(1+ep*(1-2*rand(1,1)))*feval(r_,(i-1)*h);
        for j=1:i;
                A(i,j)=16*sqrt(i-j+1)*h^(3/2);
        end;
end;
f=A\rvec';
```

B.5 Linear Algebra Scripts

B.5.1 Module 5.2

```
%.................................................................
% Projection onto a hyperplane
%.................................................................
%
        function [p]=proj(v,x,mu);
%
%.................................................................
%
% Inputs:
%      v = direction vector for hyperplane  (1xn)
%      x = vector to be projected  (1xn)
%      mu= hyperplane constant
%
% Outputs:
%      p = projected vector
%
%.................................................................
%
%      p = projection of x onto the hyperplane
%             {y: (v,y) = mu}
%
%.................................................................
       p = x + ((mu - x*v')/norm(v)^2)*v ;
```

```
%.............................................................
%
% Program to perform algebraic reconstruction technique
%
%.............................................................
%
       function [x] = art1(V,mu,x0,n);
%
%.............................................................
%
% Inputs:
%
% V  = view matrix: rows are view vectors
% mu = vector of projection values: mu(i) = proj. value of view V(i,:)
% x0 = initial approximation to solution (1xn vector)
% n  = number of complete projection cycles to perform
%
% Outputs:
%
% x  = approximate solution from ART algorithm with constraints
%.............................................................
%
% This program uses the algebraic reconstruction iterative technique,
% with projection onto the positive orthant after each iteration cycle
%
%.............................................................
%
m=size(V,1);
xold=x0;
for j=1:n
       for k=1:m;
               x=proj(V(k,:),xold,mu(k));
               xold=max(x,zeros(size(x)));
               xold=min(xold,ones(1,size(xold)));
               x=xold;
       end;
end;

%.............................................................
%
% Program for crude display of tomographic imaging
%
%.............................................................
%
       function [n]=displa(x);
%
%.............................................................
%
```

```
% Input:
%
%       x = row vector of relative shadings, size = a square integer
%
% Output:
%
%       n = square root of size of x, used for check
%.................................................................
%
 axis('equal');
 n = sqrt(size(x,2));
 xmax=max(x);
 for k=1:n;
        for l=1:n;
                [X,Y,s]=shade(l-1,n-k,x(n*(k-1)+l)/xmax);
                fill(X,Y,s); hold on;
        end;
 end;

%.................................................................
%
% Program to create arrays for shading in tomography
%
%.................................................................
%
        function [X,Y,s]=shade(i,j,p);
%
%.................................................................
%
% Inputs:
%
%       (i,j)= coordinates of southwest corner of 1x1 box to be shaded
%         p  = degree of shading, 1 = black, 0 = white
%
% Outputs:
%
%       X = matrix of x-coordinates for 'fill' command
%       Y = matrix of y-coordinates for 'fill' command
%       s = color vector for 'fill' command
%
%.................................................................
%
 X = i+[0 0 1 1 0 1 0 1];
 Y = j+[0 1 0 1 0 0 1 1];
 s = (1-p)*[1 1 1];
```

B.5.2 Module 5.3

```
%.................................................................
%    Discretization of first kind integral equation for one-
%    dimensional geophysical prospecting problem.
%.................................................................
%
     function [A,b,s,w]=Geo(n,ep);
%
%.................................................................
%
%  Inputs:
%
%    n  = number of subintervals used to discretize integral
%    ep = amplitude of random noise in right hand side
%
%  Outputs:
%
%    A  = matrix discretization of integral operator
%    b  = discretized right hand side
%    s  = vector of sampling points
%    w  = vector of values of true solution
%.................................................................
%
%    This program discretizes the right hand side for the model problem
%    in which the true mass distribution is w(x)=1.
%    Right hand side=(1-x)/((x-1)^2+1)^(1/2) + x/(x^2+1)^(1/2)
%    The mid-point rule is used to discretize the integral.
%
%.................................................................
%
%  Initializations
%
      h=1/n;     s=h*([1:n]'-0.5);    w=ones(n,1);
%
%  Integral is discretized by mid-point rule and collocation at
%  mid-points.
%
      A = h*((s*ones(1,n) - ones(n,1)*s').^2 + ones(n,n)).^(-3/2);
%
%  Discretization of right-hand side and introduction of random errors
%
    b=(ones(n,1)-s)./sqrt((s-ones(n,1)).^2 + ones(n,1));
    b=b + s./sqrt(s.^2 + ones(n,1));
    b=b+ep*(ones(n,1)-2*rand(n,1));
```

B.5.3 Module 5.5

```
%..................................................................
%
% Generates direct matrix and discretized spherical density for
% stellar tomography problem
%
%..................................................................
%
        function [A,y]=globd(R,n,f_);
%
%..................................................................
%
% Inputs:
%
%        R = radius of cluster
%        n = number of subintervals for piecewise linear approximations
%        f_= a user defined sherical density
%
% Outputs:
%
%        A = nxn matrix which discretizes the direct projection operator
%        y = n-vector discretizing the spherical density
%
%..................................................................
h=R/n;
A=zeros(n,n);
for i=1:n;
     for j=i:n;
        A(i,j)=2*h*(sqrt(j*j-(i-1)^2)-sqrt((j-1)^2-(i-1)^2));
     end;
     y(i)=feval(f_,(i-1)*h);
end;
```

Bibliography

A. Armitage, *Edmond Halley*, Nelson, London, 1966.

W. Rouse Ball, *An Essay on Newton's "Principia,"* Macmillan, New York, 1893.

B. Bollobas, ed., *Littlewood's Miscellany*, Cambridge University Press, Cambridge, 1986.

B. A. Bolt, *Inside the Earth: Evidence from Earthquakes*, Freeman, San Francisco, 1982.

J. B. Brackenridge, *The Key to Newton's Dynamics*, University of California Press, Berkeley, 1996.

W. J. Broad, Earth's rapidly spinning core is a planet within a planet, *New York Times*, December 17, 1996.

J. D. Burchfield, *Lord Kelvin and the Age of the Earth*, Science History Publications, New York, 1975.

H. R. Burger, *Exploration Geophysics of the Shallow Subsurface*, Prentice-Hall, Englewood Cliffs, NJ, 1992.

G. E. Christianson, *Edwin Hubble: Mariner of the Milky Way*, Farrar, Strauss and Giroux, New York, 1995.

A. Cook, *Edmond Halley: Charting the Heavens and the Seas*, Clarendon Press, Oxford, 1998.

A. M. Cormack, Early two-dimensional reconstruction and recent topics stemming from it, *Science* **209** (1980), pp. 1482–1486.

——— Computed tomography: some history and recent developments, *Proceedings of Symposia in Applied Mathematics* **27** (1982), pp. 35–42.

S. Drake, *Galileo Studies*, University of Michigan Press, Ann Arbor, 1970.

S. Drake and I. E. Drabkin, *Mechanics in Sixteenth-Century Italy*, University of Wisconsin Press, Madison, 1969.

D. L. Eicher, *Geologic Time*, Prentice-Hall, Englewood Cliffs, NJ, 1968.

G. Galilei, *Two New Sciences* (Elzevirs, Leyden, 1638), translated with a new introduction and notes by Stillman Drake, Second Edition, Wall and Thompson, Toronto, 1989.

G. Gamow, *Gravity*, Anchor Books, New York, 1962.

R. Gordon, G. Herman, and S. Johnson, Image reconstruction from projections, *Scientific American* **233** (October 1975), pp. 56–68.

C. W. Groetsch, *Inverse Problems in the Mathematical Sciences*, Verlag Vieweg, Braunschweig, 1993.

——— Halley's gunnery rule, *The College Mathematics Journal* **28** (1997), pp. 47–50.

M. Grosser, *The Discovery of Neptune*, Harvard University Press, Cambridge, MA, 1962.

A. Holmes, *The Age of the Earth*, Harper, New York, 1913.

D. Howse, *Nevil Maskelyne: The Seaman's Astronomer*, Cambridge University Press, Cambridge, 1989.

M. K. Hubbert, *The Theory of Ground-Water Motion* (reprint), Hafner, New York, 1969.

H. S. Jones, *John Couch Adams and the Discovery of Neptune*, Cambridge University Press, Cambridge, 1947.

A. Kirsch, *An Introduction to the Mathematical Theory of Inverse Problems*, Springer, New York, 1996.

R. Osserman, *The Poetry of the Universe*, Anchor Books, New York, 1995.

I. Peterson, Inside averages, *Science News* **129** (May 1986), pp. 300–302.

J. H. Poynting, *The Earth, Its Shape, Size, Weight, and Spin*, Cambridge University Press, Cambridge, 1913.

L. A. Shepp and J. B. Kruskal, Computerized tomography: the new medical x-ray technology, *American Mathematical Monthly* **85** (1978), pp. 420–438.

R. S. Westfall, *Never at Rest: A Biography of Isaac Newton*, Cambridge University Press, Cambridge, 1980.

S. D. Wicksell, The corpuscle problem, *Biometrika* **17** (1925), pp. 84–99.

F. W. Wintherbotham, *The Ultra Secret*, Harper and Row, New York, 1974.

Index